Undergraduate Texts in Physics

Undergraduate Texts in Physics (UTP) publishes authoritative texts covering topics encountered in a physics undergraduate syllabus. Each title in the series is suitable as an adopted text for undergraduate courses, typically containing practice problems, worked examples, chapter summaries, and suggestions for further reading. UTP titles should provide an exceptionally clear and concise treatment of a subject at undergraduate level, usually based on a successful lecture course. Core and elective subjects are considered for inclusion in UTP.

UTP books will be ideal candidates for course adoption, providing lecturers with a firm basis for development of lecture series, and students with an essential reference for their studies and beyond.

More information about this series at http://www.springer.com/series/15593

Byron P. Roe

Probability and Statistics in the Physical Sciences

Third Edition

 Springer

Byron P. Roe
Randall Laboratory
University of Michigan
Ann Arbor, MI, USA

ISSN 2510-411X   ISSN 2510-4128   (electronic)
Undergraduate Texts in Physics
ISBN 978-3-030-53693-0   ISBN 978-3-030-53694-7   (eBook)
https://doi.org/10.1007/978-3-030-53694-7

This Springer imprint is published by the registered company Springer Nature Switzerland AG
The registered company address is: Gewerbestrasse 11, 6330 Cham, Switzerland

Preface

This book is meant to be a practical introduction to the use of probability and statistics in experimental physics for advanced undergraduate students and for graduate students. I have tried to emphasize the areas I have found to be useful for the experimental sciences. Except for the first two chapters, the emphasis is on applications and understanding. I have omitted proofs of formal theorems in the interests of brevity unless I felt the proof added to one's intuition in understanding and applying the theorem. Since, however, this is a field in which there are often a number of misunderstandings, it is necessary to state some things with reasonable precision. I have tried to do this when necessary.

I assume the student is familiar with partial derivatives and with elementary matrix manipulation. A computer also is a needed tool for probability and statistics in experimental physics and will be used in many of the homework exercises. A large library of important routines for statistical use are available at CERN, the European Organization for Nuclear Research (https://home.cern). Given the path to CERN, CERN/root contains this library and has an introduction. Programs are usually written in C^{++}. Among many other things the library includes pseudorandom number routines, histogram routines, minimizing routines, neural net and boosted decision tree routines, and matrix manipulation.

In the interactive mode, the Mathematica software system, available commercially, is also extremely useful.

Any of these systems will require an initial learning effort. However, the reward in terms of problem-solving capability is very great and the investment in time well spent.

In some of the exercises I will provide some hints on the use of the CERN system. A solutions manual is now available.

For the Second Edition, I made a number of changes and additions. Among the most important of these are a new chapter on queueing theory and a discussion of the Feldman–Cousins unified method for estimating confidence intervals. Expositions have been added on the fitting of weighted events, the fitting of events with errors in x and y, and a number of other topics.

For the Third Edition, I have added a new chapter on Event Characteristics, including introductions to neural nets, convolutional neural nets (deep learning), adversarial neural nets and boosted decision trees. Major changes and additions have been made throughout the book.

Ann Arbor, USA   Byron P. Roe

Acknowledgements

Many people have contributed to the preparation of this Third Edition. I could not have completed it without their help.

Jean Hue helped enormously in translating some of the old answers to exercises into more modern C++ language. He also carefully read the text from the unique perspective of an undergraduate student for whom some of the material is new. He suggested many very useful changes to clarify or expand sections of the text.

Dr. Johnathon Jordan in our Physics Department has very greatly helped me to understand the new forms for the extensive CERN library using "root" both to use their sophisticated programs to fit experimental data to theory and to use it to make plots and histograms.

Dr. Charles Antonelli, in the LSA Research Computing Support group, helped me to solve some TeX problems to obtain references by chapter.

Brandon Case in the Physics Department Computing Support group helped me to install the GSL (GNU Scientific Library).

Finally, my colleague, Professor of Physics, Leonard Sander read through my draft and made numerous very useful suggestions.

Contents

Chapter 1
Basic Probability Concepts

Abstract Central to our study are three critical concepts: *randomness*, *probability*, and *a priori probability*. In this short chapter, we will discuss these terms. Most of the applications in this text deal with classical not quantum phenomena. In a deterministic universe, how can we understand these concepts? Probability is a very subtle concept. We feel we intuitively understand it. Mathematically, probability problems are easily defined. Yet when we try to obtain a precise physical definition, we find the concept often slips through our grasp.

1.1 Discussion of Randomness, Probability, and a Priori Probability

From the point of view of pure mathematics, there is no problem. We will deal with the properties of a function, $F(x')$, which changes monotonically from 0 to 1 (continuously or discontinuously) as x' goes from negative to positive infinity. $F(x')$ is called the distribution function. It is said to be "the probability that x is less than or equal to x'." The derivative, $f(x') = dF(x)/dx \mid_{x'}$, is called the probability density function. Where it exists $f(x')dx'$ is described as the "probability of x being between x' and $x' + dx'$." Generally the $\int_{x=x_1}^{x=x_2} dF(x)$ is defined as "the probability that x is between x_1 and x_2." The problems arise when, somehow, we wish to connect this kind of probability with the real world.

What is randomness? Coin tosses or throws of dice are basically classical, not quantum mechanical, phenomena. How can we have randomness in a deterministic classical system? Suppose we build a coin tossing machine, which tosses a coin over and over again. If we examine the springs and pivots carefully enough, can we predict what would be the sequence of heads and tails in 100 throws?

Starting in about the mid-1960s, we have finally been able to come to grips with this question and to see that in practice we cannot make this kind of prediction. We can now see how randomness enters into deterministic physics. Ford has written a very nice article on this subject (Ford 1983). I will summarize some of the main concepts.

© Springer Nature Switzerland AG 2020

B. P. Roe, *Probability and Statistics in the Physical Sciences*,
Undergraduate Texts in Physics,
https://doi.org/10.1007/978-3-030-53694-7_1

Imagine that we write a computer program to predict the results of a sequence of experiments such as a string of coin tosses. Suppose the law is put in the form of some sort of difference equation with some sort of initial conditions. As the string of tosses gets longer and longer, the difference equation remains the same, but the initial conditions need to be specified more and more accurately. Therefore, the length of the program can come to be dominated by the number of bits needed for the initial conditions. If the number of bits needed for the program including the initial conditions is more than the number of output bits, then the program is of limited efficiency for predicting results and for organizing experience. We could save space by just keeping the data. If the ratio of program size to output bits does not approach zero as the output string we wish to produce gets longer and longer, then the solution is chaotic. For reasonable predictability, we need to ask that the number of bits in the computer program should be smaller than the number of bits we are trying to predict.

Next we turn to the physical equation. Many equations have some solutions that tend to grow quickly, perhaps exponentially. It turns out, in fact, that most physical equations have this sort of solution. If there is an exponentially increasing piece in the specific solution in which we are interested, then the initial conditions need to be specified with extreme accuracy, and we run into the problem stated above; the initial conditions eventually use more bits than the number of bits describing the output string of heads and tails (n versus $\log n$). Although we might predict the results of the first few tosses, the specification of initial conditions quickly goes beyond our capability. If we increase the number of binary bits in the initial conditions by n, we only increase the length of the predictable string by $\log n$. Thus, effectively, we lose predictive power and we are justified in defining these as random processes. People have stated that a butterfly flapping her wings in China may cause a storm in California. These kinds of studies have grown into a whole mathematical subdiscipline known as chaos theory (Strogatz 2015). Hence, random processes come out of deterministic equations. Using quantum mechanics it is possible to get truly random numbers, but, usually, we deal with pseudo-random numbers.

The second concept we wish to discuss is probability. Perhaps the most popular definition of probability is that it is the "limit of relative frequencies." For example, suppose we have many identical independent random tosses of a die. Let n equal the number of times 4 comes up and N equal the number of trials. Imagine that we do many experiments, i.e., many sets of N trials each. Then we find that for most of these experiments, n/N approaches a limit as N approaches infinity. We take this limit to be the probability of 4 coming up on the die.

There are some problems with the above definition. The first of these concerns is the notion of identical trials. Practically, it is not possible to have identical trials. In the above example, the die will wear out, the corners will become rounded. We certainly can minimize this, for example, by using a die cut from a single large diamond and tossing it onto silk batting. However, we cannot eliminate it. Many thermodynamical or statistical mechanics applications are afflicted with an extreme form of this problem. They often begin with "Imagine we had a set of universes." However, we can only examine the one in which we live.

Another problem will arise as we go further into the formalism. We will find that for any N, there is some probability that n/N can be arbitrary (even for a string of random experiments) and, therefore, perhaps far away from what one expects. The probability of this occurring, indeed, falls to zero as N increases, but is present for any finite N. Hence, one must say that the probability is usually or probably the limit of relative frequencies. Hence, the definition becomes if not circular, at least spiral.

In trying to define probability above, we used the concept of independent trials. Even this term has come under fire. If a "good" die comes up with the 4 side uppermost 15 times, is the probability one-sixth for a 4 the 16th time or less? Does nature get tired of long strings and does the probability of 4 approach 0 (or some other value) after a long string of 4's? The German philosopher K. Marbe (quoted in Feller's book (Feller 1950) on probability) introduced this latter assumption into his philosophic system, i.e., endowed nature with a memory. This is a perfectly consistent philosophical assumption. *Experimentally*, it is wrong. The probability does remain the same for the 16th trial as it was for the first trial. (We have probably all seen philosophical descendents of Marbe in any games of chance we have played. "I've lost three times in a row. Surely this time I'm bound to win.")

As we see, the limiting relative frequency definition of probability is too simplistic. Furthermore, it mixes concepts. The formal definition of probability in terms of the distribution function includes the concept of limiting relative frequency. As we will see later, it specifies in what way the limit is approached and addresses quantitatively the special strings which stay far from the expected limit. Within the formal theory, we do not need a further definition of probability. The real question is whether this theory applies to coin tosses. That is an empirical question as we have already seen. It depends on whether the trials are independent and the equations chaotic, and, finally, whether the predicted results agree with the observation: *Probability theory is mathematics. The set of applications is physics.*

A priori probability is the last of our terms. It is the probability of an occurrence estimated before doing the experiment. For instance, one-sixth would be the a priori probability for the 4 side to come up for a "good" die. If the relative frequency comes out far from one-sixth, we would look for a physical cause. Perhaps this is not a "good" die. Then we would say one-sixth was a poor guess and the a posteriori probability (i.e., the probability after the experiment) of having a 4 come up would be taken as something other than one-sixth. Thus, a priori probability is what we *thought* the probability was. A priori probability was often criticized by physicists in the 1950s. However, there are important places where it is quite useful and some of these will be discussed in this text.

It is amusing to note that the a priori probability does not even have to be one of the possible a posteriori values. Suppose we have a box containing a large number of dice of a rather peculiar kind. On each die, all faces are the same and equal numbers of dice containing each of the six possible numbers are present. We randomly choose a die without looking at it. The a priori probability of tossing it and having it land with a 4 up is one-sixth, the fraction of dice with 4's. However, if we toss the die and a 4 is uppermost, the a posteriori probability of rolling that die and getting a 4 is one. In this instance, the possible a posteriori results are zero or one, not one-sixth.

We have now defined both the mathematical concepts and the physical concepts needed to proceed with our probability studies. The distribution and density functions are the basic tools we use, but to use them we must analyze each physical situation to make sure that it meets the criteria described earlier. These criteria include randomness and, depending on the situation, independence and identical trials.

References

Feller W (1950) Probability theory and its applications, vol 1. Wiley, New York
Ford J (1983) How random is a coin toss? Phys Today 40(4)
Strogatz S (2015) Nonlinear dynamics and chaos, with applications to physics, biology, chemistry, and engineering, 2nd edn. Westview Press, Boulder

Chapter 2
Some Initial Definitions

Abstract In this chapter, we will introduce some terms to give us a common language. We will be dealing for the most part with properties of a nondecreasing function of x, which goes from 0 at the lower limit to 1 at the upper limit of x. We will define density function, distribution function, moments of a distribution, independence and correlations of various probabilities, and various measures of the center of a distribution.

2.1 Basic Concepts

In order to try to put some flesh on these bare bones, we will assume intuitive ideas to make a first simple connection with the world.

Sample space: This is the space of all possible outcomes of an experiment.

Random variable: This is a function defined on the sample space. For example, if you measure x, then x^2 is a random variable.

Distribution function: We define this in one dimension first. Suppose the sample space is one-dimensional (x) space. The distribution function $F(x')$ is the probability that when you measure a value of x, it is less than or equal to x'. $F(-\infty) = 0$. $F(+\infty) = 1$. F is a nondecreasing function of x. It can only stay constant or increase as x increases. It can change continuously or discontinuously. The distribution function is sometimes called the cumulative distribution function.

Discrete probability: A discrete variable is one with a countable number of distinct values. For a discrete variable sample space we define P_r as the probability that the outcome is r. The sum over all r of P_r is 1.

Density function: This is defined if we have a continuous variable sample space. The probability density function (pdf) is $f(x) = dF/dx$ in one dimension. It is sometimes also called the frequency function, or the differential probability function. $f(x')dx'$ is the probability that x lies between x' and $x' + dx'$. The integral over all x' of $f(x')$ is 1. Note that whereas F is dimensionless, f has the dimension of x^{-1}.

Multidimensional extensions of the above definitions are straightforward. P_{rs} is the two-dimensional discrete variable probability function. It is the probability that

© Springer Nature Switzerland AG 2020

B. P. Roe, *Probability and Statistics in the Physical Sciences*,
Undergraduate Texts in Physics,
https://doi.org/10.1007/978-3-030-53694-7_2

both r and s occur. $F(x_1', x_2')$ is the probability that x_1 is less than or equal to x_1' and at the same time x_2 is less than or equal to x_2':

$$f(x_1, x_2) = \frac{\partial^2 F}{\partial x_1 \partial x_2}. \tag{2.1}$$

Marginal probability: For discrete probability, this is the probability that r occurs regardless of s. P_r is sum over all s of P_{rs}. For continuous variables,

$$F_1(x_1) = F(x_1, \infty), \qquad f_1(x_1) = \frac{dF_1}{dx_1}. \tag{2.2}$$

The 1 in F_1 indicates that it is the first variable that is retained.

Conditional probability: This is the probability that r occurs given that $s = s_0$. For the discrete case,

$$P\{r|s_0\} = \frac{P_{rs_0}}{P_{s_0}}. \tag{2.3}$$

For the continuous variable case,

$$f_1(x_1|x_2) = \frac{f(x_1, x_2)}{f_2(x_2)}. \tag{2.4}$$

Here, $f_2(x_2)$ is the pdf for x_2 regardless of x_1.

Expectation value: Let $g(x)$ be a random variable as defined above. The expectation value of g is the average value of g expected in the experiment. Precisely, it is defined as

$$\mathbb{E}\{g\} \equiv \bar{g} = \int_{x=-\infty}^{x=+\infty} g \, dF = \int_{-\infty}^{+\infty} gf \, dx, \tag{2.5}$$

for x a continuous variable and $= \sum_r g_r P_r$ if we have a discrete variable.

For the conditional probability that x_1 occurs given that x_2 has occurred, $P\{x_1|x_2\}$, we can show,

$$\mathbb{E}\{\mathbb{E}\{x_1|x_2\}\} = \mathbb{E}\{x_1\}. \tag{2.6}$$

This follows since

$$\mathbb{E}\{x_1|x_2\} = \int_{-\infty}^{\infty} x_1 f\{x_1|x_2\} \, dx_1,$$

$$\mathbb{E}\{\mathbb{E}\{x_1|x_2\}\} = \int_{-\infty}^{\infty} \int_{-\infty}^{\infty} x_1 \frac{f(x_1, x_2)}{f_2(x_2)} f_2(x_2) \, dx_1 \, dx_2$$

$$= \int_{-\infty}^{\infty} x_1 \int_{-\infty}^{\infty} f(x_1, x_2) \, dx_2 \, dx_1$$

$$= \int_{-\infty}^{\infty} x_1 f_1(x_1) \, dx_1 = \mathbb{E}\{x_1\}.$$

Probability moments: These are expectation values of various powers of random variables. Let $g = x^n$.

$$\text{Then } \overline{x} = mean\ value\ of\ x = \text{first moment} = m,$$
$$\overline{x^2} = \text{second moment} \ldots \tag{2.7}$$

We can also define central moments. These are moments about the mean value of x (i.e., $x - m$). Let $g = (x - m)^n$:

$$\overline{(x - m)} = 0, \tag{2.8}$$
$$\overline{(x - m)^2} = \text{second central moment} = \sigma^2 = variance. \tag{2.9}$$

$\sigma (= \sqrt{\sigma^2})$ is called the *standard deviation*.

$$\overline{(x - m)^n} = \mu_n. \tag{2.10}$$

Some functions of these central moments are also sometimes used to categorize distributions:

$$\gamma_1 = \mu_3/\sigma^3 = \text{coefficient of skewness.} \tag{2.11}$$
$$\gamma_2 = \mu_4/\sigma^4 - 3 = \text{kurtosis } or \text{ coefficient of excess.} \tag{2.12}$$

Before discussing some of these, let us prove an important theorem.

$$\text{Consider } \sigma^2 = \overline{(x - m)^2}$$
$$= \overline{(x^2 - 2xm + m^2)} \tag{2.13}$$
$$= \overline{x^2} - 2m\overline{x} + m^2.$$

Note that $2m$ is a constant and $\overline{x} = m$. Thus, our theorem is

$$\sigma^2 = \overline{x^2} - (\overline{x})^2. \tag{2.14}$$

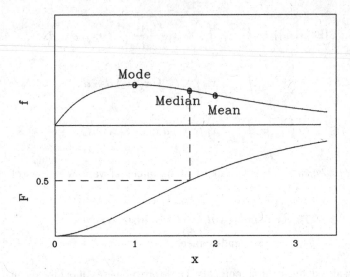

Fig. 2.1 Illustration of mode, median, and mean of a distribution

We have defined the mean, m, in Eq. 2.7. This is one measure of an average or central point of a distribution. Another might be the peak of the density function. This is called the *mode*. Still another is the middle value of the distribution, the *median*. This occurs when the distribution function is 1/2. These measures are illustrated in Fig. 2.1.

The quantity σ, the standard deviation we discussed earlier, is a measure of the width of the distribution, since it measures how far individual trials vary from the mean. γ_1, the coefficient of skewness, is a quantity that is 0 if the density function is symmetric about the mean value. γ_2, the kurtosis, measures the deviation from a normal or Gaussian distribution, which we will discuss shortly. If the kurtosis is positive, the tails of the distribution are wider than they are for a normal distribution.

Dependence and independence: Two variables are independent if and only if

$$F(x_1, \ x_2) = F(x_1) * F(x_2). \tag{2.15}$$

They are then said to be *uncorrelated*. If this relation is not satisfied, they are said to be dependent and there are (usually) correlations between x_1 and x_2.

Correlation coefficient: The covariance between two variables Cov_{12}, is defined as

$$Cov_{12} = \overline{(x_1 - \overline{x}_1) * (x_2 - \overline{x}_2)}. \tag{2.16}$$

The correlation coefficient between two variables is defined as

$$C_{12} = \frac{Cov_{12}}{\sqrt{\sigma_1^2 \sigma_2^2}} \tag{2.17}$$

By an argument similar to that leading to Eq. 2.14, we see that

$$Cov_{12} = \overline{x_1 x_2} - \overline{x_1}\,\overline{x_2}. \tag{2.18}$$

We also note that

$$\text{variance}(x_1 + x_2) = \text{variance}(x_1) + \text{variance}(x_2) + 2 \times \text{covariance}(x_1, x_2). \tag{2.19}$$

If two variables are uncorrelated, then $C_{12} = 0$. However, $C_{12} = 0$ does not necessarily imply two variables are independent.

Let us give an example of this. Suppose we toss two dice. We define three random variables:

Let r be 1 if die 1 has an odd face up and 0 otherwise,

  s be 1 if die 2 has an odd face up and 0 otherwise,

   t be 1 if an odd sum is up and 0 otherwise.

These events are pairwise independent. Let P_r be the probability that $r = 1$, and so on. Then:

$$\sum_t P_{rst} = P_r P_s, \qquad \sum_s P_{rst} = P_r P_t, \qquad \sum_r P_{rst} = P_s P_t. \tag{2.20}$$

Amazingly enough, this does not imply that the variables are independent, i.e., uncorrelated. If r and s both occur, t is excluded. For independence, we need

$$P_{rst} = P_r P_s P_t. \tag{2.21}$$

Pairwise independence is not enough. This example was due to Feller (1950).

As you can see from the above, you have to be very careful how things are worded in probability studies. Here is another example of the general subtlety in probability considerations. Imagine a birth is equally likely to be a boy or a girl.

Case 1: Suppose we are given that a family has two children and at least one is a boy. What is the probability that they are both boys? The answer is 1/3. The various choices are (older one listed first): boy–boy, boy–girl, girl–boy, girl–girl. Only the first three of these cases have at least one boy and, in two out of the three, the other child is a girl.

Case 2: Choose a boy at random. Suppose he comes from a family of two children. What is the probability his sibling is a boy? The answer here is 1/2!!

For case 1, we looked through a card file of families. For case 2, we looked through a card file of boys. There were two boys in the first family and, therefore, it was counted twice.

We have tried in this chapter to define some of the basic mathematical terms you will need to work with probability and to give some simple initial examples indicating how to apply these concepts in practice. In these examples, we have tried to illustrate both the power and the subtlety involved in this field.

2.2 Worked Problems

WP2.1 Suppose particles can fall randomly in a square of side d. Find $\bar{x}$ and σ^2.
See Fig. 2.2.

Answer:

This problem corresponds to having the density function $f(x)$ a constant for
$0 < x < d$ and 0 for $x < 0$ or $x > d$ (uniform distribution). Since we must have
$\int_{-\infty}^{\infty} f \, dx = 1$, the constant must be $1/d$. Hence,

$$\bar{x} = \int_0^d (1/d)x \, dx = d/2.$$

$$\overline{x^2} = \int_0^d (1/d)x^2 \, dx = d^2/3.$$

Since $\sigma^2 = \overline{x^2} - \bar{x}^2$, we have

$$\sigma^2 = d^2/3 - (d/2)^2 = d^2(1/3 - 1/4) = d^2/12.$$

Thus, $\sigma = d/\sqrt{12}$.

WP2.2 Suppose the square above is tipped by $45°$. Find $\bar{x}$ and σ^2. See Fig. 2.3.

Answer:

Here $f(x) \propto 1 - |x/(\sqrt{2}d/2)|$, for x between $\pm d/\sqrt{2}$ and 0 otherwise, where
the diagonal $= \sqrt{2}d$. Thus, $f(x) = C(1 - |x|\sqrt{2}/d)$. Since $\int_{-\infty}^{\infty} f \, dx = 1$, we
have

Fig. 2.2 Square of side d

Fig. 2.3 Square of side d tipped by 45° with respect to the x axis

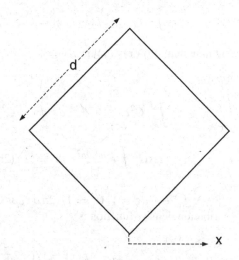

$$2C \int_0^{d/\sqrt{2}} (1 - x\sqrt{2}/d)\, dx = 1 = 2C(d/\sqrt{2} - d/(2\sqrt{2})).$$

where the factor of 2 occurs since x can be positive or negative. Thus, $C = \sqrt{2}/d$.

$$f(x) = (\sqrt{2}/d)(1 - |x|\sqrt{2}/d).$$

This implies that $\bar{x} = 0$ since x is an odd function. Thus,

$$\sigma^2 = \overline{x^2} = 2 \int_0^{d/\sqrt{2}} \left(\sqrt{2}/d\right)(1 - x\sqrt{2}/d)x^2\, dx$$
$$= \left(2\sqrt{2}/d\right)(d/\sqrt{2})^3(1/3 - 1/4) = d^2/12.$$

This is the same variance we obtained in WP2.1. Although x_{max} was bigger in this problem than in the first problem, it was compensated for by the fact that more of the area was at low x. (Note the similarity to moment of inertia calculations.)

WP2.3 The normal or Gaussian distribution has

$$f \propto e^{-x^2/2\sigma^2}, \quad \text{i.e., } f = Ce^{-x^2/2\sigma^2}, \tag{2.22}$$

with σ a fixed number. Find C. There is a trick here since $\int_{-\infty}^{\infty} f\, dx$ cannot be written in a simple closed form. Consider

$$f(x)f(y) = C^2 e^{-x^2/2\sigma^2} e^{-y^2/2\sigma^2}, \tag{2.23}$$

and note that $\iint f(x)f(y)\,dx\,dy = 1$.

Answer:

$$\iint C^2 e^{-(x^2+y^2)/2\sigma^2} dx\,dy = C^2 \iint e^{-r^2/2\sigma^2} r\,dr\,d\theta = 1. \tag{2.24}$$

$$2\pi C^2 \int e^{-r^2/2\sigma^2} r\,dr = (2\pi(2\sigma^2)C^2/2)\int e^{-z}\,dz = 1. \tag{2.25}$$

Since $\int_0^\infty e^{-z}\,dz = 1$, $C^2 = 1/(2\pi\sigma^2)$ or $C = 1/(\sqrt{2\pi}\sigma)$. Thus, the normal distribution density function is

$$f(x) = 1/(\sqrt{2\pi}\sigma)e^{-x^2/2\sigma^2}.$$

The trick that we have used here is useful in generating on a computer a set of pseudorandom numbers distributed in a normal distribution as we will see in Chap. 8.

2.3 Exercises

2.1 A 1-cm-long matchstick is dropped randomly onto a piece of paper with parallel lines spaced 2 cm apart marked on it. You observe the fraction of the time that a matchstick intersects a line. Show that from this experiment you can determine the value of π. (This problem is known as the Buffon's Needle problem.)

2.2 Let $\alpha_n = \overline{x^n}$ be the nth moment of a distribution, μ_n be the nth central moment of the distribution, and $m = \overline{x}$ be the mean of the distribution. Show that

$$\mu_3 = \alpha_3 - 3m\alpha_2 + 2m^3.$$

2.3 Suppose one has two independent radioactive sources. Disintegrations from each are counted by separate scintillators with expected input rates N_1 and N_2. Ignore the dead times of the array. Suppose one is measuring the coincidence rate, that is, asking how often the two scintillators are excited simultaneously. If the circuitry used will count two signals as being simultaneous if they are less than a time τ apart, find the measured coincidence rate. Imagine $N_1\tau \ll 1$ and $N_2\tau \ll 1$. This calculation is useful as a calculation of background in experiments in which we are trying to measure single events that cause both counters to count simultaneously.

2.4 A coin is tossed until for the first time the same result appears twice in succession. To every possible pattern of n tosses, attribute probability $1/2^n$. Describe the sample space. Find the probability of the following events: (a) the experiment ends before the sixth toss, (b) an *even* number of tosses is required. Hint: How many points are there in the sample space for a given number n?

Reference

Feller W (1950) Probability theory and its applications, vol 1. Wiley, New York

Chapter 3
Some Results Independent of Specific Distributions

Abstract We could start out and derive some of the standard probability distributions. However, some very important and deep results are independent of individual distributions. It is very easy to think that many results are true for normal distributions only when in fact they are generally true. We will discuss multiple scattering, the $\sqrt{N}$ law, propagation of errors, errors when changing variables, correlation coefficients, and some generally useful inequalities.

3.1 Multiple Scattering and the Root N Law

In this section we will deal with a general result of great use in practice. Paradoxically, we begin with a specific example, multiple scattering of a charged particle passing through matter. However, the square root N law we obtain is very general and appears in many areas of physics.

Suppose a charged particle is traveling in the x direction through a lattice of atoms of total length L and is scattered by a number of the atoms it passes. Consider the two-dimensional case to begin with. See Fig. 3.1 Let i be the site of the ith atom along x, and θ_{yi} be the angle through which that atom is scattered.

$$\theta_y = \sum \theta_{yi}, \tag{3.1}$$

$$\overline{\theta_y} = 0 \text{ assuming the scattering is symmetric} \tag{3.2}$$

$$\overline{\theta_y^2} = \overline{\left(\sum \theta_{yi}\right)^2} = \overline{\sum \theta_{yi}^2 + \sum_{i \neq j} \theta_{yi}\theta_{yj}}. \tag{3.3}$$

The basic assumption is that each scattering is independent! Then θ_{yi} is independent of θ_{yj} for $j \neq i$. If energy loss is ignored, the scattering is independent of x. Therefore,

© Springer Nature Switzerland AG 2020

B. P. Roe, *Probability and Statistics in the Physical Sciences*,
Undergraduate Texts in Physics,
https://doi.org/10.1007/978-3-030-53694-7_3

Fig. 3.1 Particle initially
traveling parallel to the x
axis undergoing multiple
scattering in a
two-dimensional problem.
After going a horizontal
distance L, it has moved y
units vertically and is
traveling at an angle θ_y with
respect to the x axis

$$\sum_{i \neq j} \overline{\theta_{yi}\theta_{yj}} = 0, \tag{3.4}$$

$$\overline{\theta_y^2} = \sum \overline{\theta_{yi}^2}, \tag{3.5}$$

$$\overline{\theta_y^2} = N\overline{\theta_{yi}^2}, \tag{3.6}$$

where N = the number of scatterings. Thus $\sqrt{\overline{\theta_y^2}} \propto \sqrt{N}$.

In integral form, we now have

$$\overline{\theta_y^2} = \iint \theta_y^2 P(\theta_y,\ x)\ dx\ d\theta_y. \tag{3.7}$$

$P(\theta_y,\ x)\ dx\ d\theta_y$ is the probability that within dx of x there is a scattering within $d\theta_y$
of θ_y. $\overline{\theta_i^2} = \int \theta_{yi}^2 P(\theta_{yi}, x) d\theta_{yi} dx$. Since the average scattering is independent of x, we
have the important result that the average number of collisions in a distance L is
$N \propto L$.

$$\overline{\theta_y^2} \propto L \quad \text{or} \quad \sqrt{\overline{\theta_y^2}} \propto \sqrt{L}. \tag{3.8}$$

This has followed from the independence of collisions only. The quantity $\sqrt{\overline{\theta_y^2}}$
is called the root mean square value of θ_y, or r.m.s. value. Our results show that the
r.m.s. value of θ_y is proportional to $\sqrt{L}$ or $\sqrt{N}$.

This basic result also appears in diffusion and many other physics processes. It is
known as the "random walk" or "drunkard's walk" problem. This problem was first
posed by Pearson and a solution found by Rayleigh (Einstein 1905). In physics, it
is best known by the formulation of Einstein in connection with Brownian motion
(Pearson 1905). Suppose a drunkard walks along a street and swings around each
lamppost he finds, going randomly forward or backward from each one. Then after

N lampposts, his r.m.s. distance from his starting point will be $\sqrt{N}$ lampposts. We will also see that this $\sqrt{N}$ factor enters into probability error estimates for the same basic reason.

Brownian motion is the chaotic motion of tiny particles suspended in a fluid as they are bombarded by the molecules of the fluid. It was the first application of these concepts in physics. The study of Brownian motion enabled us to make the first estimate of the size of Avogadro's number.

Diffusion may be considered as a type of multiple scattering in which the scattering is isotropic at each collision. For the case of a labeled molecule diffusing through a gas, our results imply that $\overline{z^2} = N\overline{d_z^2}$, where $\overline{d_z^2}$ is the mean square z displacement per step. Since $d_z = v_z t$, we have $\overline{d_z^2} = \overline{v_z^2 t^2}$. From symmetry, we know that $\overline{v_x^2} = \overline{v_y^2} = \overline{v_z^2}$ and, therefore, $\overline{v_z^2} = \frac{1}{3}\overline{v^2}$. If the mean time between collisions is τ, then N collisions take a mean time $T = N\tau$ and we have

$$\overline{z^2(T)} = \tfrac{1}{3}\overline{v^2}t^2 T/\tau. \tag{3.9}$$

This then tells us the rate of diffusion. (From statistical mechanics considerations, one can show from these results that the diffusion constant D is $1/3\overline{v^2}\tau$.)

Let us go back to the problem at hand. Using the Rutherford law, one can derive that for projected angles, i.e., angles projected onto the x-y plane, we have

$$\sqrt{\overline{\theta_y^2}} = 15.2 \text{ MeV} \left((pc\beta)^{-1} \sqrt{\frac{L}{L_R}} \right), \tag{3.10}$$

where L_R = radiation length = characteristic length for a given material and is $\sim Z^2/(A*\text{density})$ and $\beta = v/c$.

How do we get to three-dimensional angles, i.e., the full scattered angles not the projected ones? Suppose that the angles we are dealing with are small. Then to a good approximation,

$$\theta_{3D}^2 = \theta_y^2 + \theta_z^2. \tag{3.11}$$

Hence,

$$\theta_{3D}^2 = \overline{\theta_y^2} + \overline{\theta_z^2} = 2\overline{\theta_y^2}. \tag{3.12}$$

Next let us calculate $\overline{y^2}$.

$$y = \sum (L - x_i)\theta_{yi}, \tag{3.13}$$

$$\bar{y} = 0, \tag{3.14}$$

$$\overline{y^2} = \sum (L - x_i)^2 \theta_{yi}^2 + \sum_{i \neq j} (L - x_i)(L - x_j)\theta_{yi}\theta_{yj}. \tag{3.15}$$

The second term is again 0 if the scatterings are independent. We go into the integral form

$$\overline{y^2} = \iint (L - x)^2 \theta_y^2 P(\theta_y, x) \, d\theta_y \, dx. \tag{3.16}$$

Again we ignore energy loss. We then have

$$\overline{y^2} = \frac{L^2}{3} \overline{\theta_y^2}. \tag{3.17}$$

For three-dimensional displacements, we have $r^2 = y^2 + z^2$ and

$$\overline{r^2} = \frac{L^2}{3} \overline{\theta_{3D}^2}. \tag{3.18}$$

Finally, consider the correlation between y and θ, $\overline{y\theta}$. This quantity would be 0 if y and θ were uncorrelated, since $\overline{y}$ and $\overline{\theta}$ are both 0.

$$\overline{y\theta} = \iint (L - x) \theta_y^2 P(\theta_y, x) \, d\theta_y \, dx, \tag{3.19}$$

$$\overline{y\theta} = \frac{L}{2} \overline{\theta_y^2}. \tag{3.20}$$

3.2 Propagation of Errors; Errors When Changing Variables

We will continue to examine results that do not depend on specific probability distributions. Propagation of errors and finding errors when changing variables are two important instances of this that we will examine in this section. We start with propagation of errors.

Suppose we measure x, which has mean value $\overline{x}$ and standard deviation σ_x, and y, which has mean value $\overline{y}$ and standard deviation σ_y. Suppose we have a function $G(x, y)$ and wish to determine the variance of G, i.e., propagate the errors in x and y to G. (Since G is then a function defined on our sample space, it is, by our definition in Chap. 2, called a random variable.)

$$G(x, y) \cong G(\overline{x}, \overline{y}) + \frac{\partial G}{\partial x}\Big|_{\overline{x}, \overline{y}} (x - \overline{x}) + \frac{\partial G}{\partial y}\Big|_{\overline{x}, \overline{y}} (y - \overline{y}). \tag{3.21}$$

We assume that the errors are "small," by which we mean that we ignore the higher-order terms in the above expansion. Then

$$\overline{G} = G(\overline{x}, \overline{y}), \tag{3.22}$$

$$\sigma_G^2 = \overline{(G - \overline{G})^2} = \overline{\left(\frac{\partial G}{\partial x}\right)^2 (x - \overline{x})^2 + \left(\frac{\partial G}{\partial y}\right)^2 (y - \overline{y})^2}$$

$$\overline{+2\frac{\partial G}{\partial x}\frac{\partial G}{\partial y}(x - \overline{x})(y - \overline{y}).} \tag{3.23}$$

If x and y are independent, we have

$$\sigma_G^2 = \left(\frac{\partial G}{\partial x}\right)^2 \sigma_x^2 + \left(\frac{\partial G}{\partial y}\right)^2 \sigma_y^2. \tag{3.24}$$

Often, however, x and y will not be independent and the correlation, i.e., the term involving $(x - \overline{x})(y - \overline{y})$, must be included.

For n variables $x_1, x_2, \ldots, x_n$, the above procedure generalizes and we have

$$\sigma_G^2 = \sum_{i,j=1}^{n} \frac{\partial G}{\partial x_i}\Big|_{\overline{x}} \frac{\partial G}{\partial x_j}\Big|_{\overline{x}} \overline{(x_i - \overline{x}_i)(x_j - \overline{x}_j)}. \tag{3.25}$$

The quantity $\overline{(x_i - \overline{x}_i)(x_j - \overline{x}_j)}$ is C_{ij}, the correlation coefficient.

A simple example of the above formalism is the case in which $G = x + y$ and the errors are independent. For this case, $\sigma_G^2 = \sigma_x^2 + \sigma_y^2$, which is the familiar relation for adding errors in quadrature.

For another example, consider the multiply scattered particle in two dimensions shown in Fig. 3.2 and again assume the angles are small. Here we have (setting $\theta_y = \theta$)

$$Y = y + D\theta,$$

$$\sigma_Y^2 = \left(\frac{\partial Y}{\partial y}\right)^2 \sigma_y^2 + \left(\frac{\partial Y}{\partial \theta}\right)^2 \sigma_\theta^2 + 2\frac{\partial Y}{\partial y}\frac{\partial Y}{\partial \theta}\overline{y\theta} \tag{3.26}$$

$$\sigma_Y^2 = \sigma_\theta^2 \left(\frac{L^2}{3} + D^2 + LD\right).$$

where we have used the relations shown in Sect. 3.1. Note that the correlation is essential in this example. Next let us consider changes of variables. Suppose we wish to change variables from $x_1, x_2, \ldots, x_n$ to $y_1, y_2, \ldots, y_n$, where $y_k = y_k(x_1, x_2, \ldots, x_n)$. It is easy to see from our previous discussion of G that

Fig. 3.2 Particle initially
traveling parallel to the x
axis undergoing multiple
scattering in a target in a
two-dimensional problem.
After going a horizontal
distance L in the target and D
afterward, it has moved Y
units away from the x axis
and is traveling at a small
angle θ_y to that axis

$$\sigma_{y_k}^2 = \sigma_G^2 = \sum_{i,j=1}^{n} \frac{\partial y_k}{\partial x_i} \frac{\partial y_k}{\partial x_j} C_{x_i x_j}. \tag{3.27}$$

Similarly

$$C_{y_l y_k} = \sum_{i,j=1}^{n} \frac{\partial y_l}{\partial x_i} \frac{\partial y_k}{\partial x_j} C_{x_i x_j}. \tag{3.28}$$

3.3 Some Useful Inequalities

In this section, we consider some useful probability inequalities. The first one is
known as the **Bonferroni inequality** and states

$$P\{EF\} \geq P\{E\} + P\{F\} - 1. \tag{3.29}$$

The Bonferroni inequality gives us a minimum value for the probability of the simultaneous occurrence of two random events. Suppose $P\{F\} \geq P\{E\}$. Then this inequality is easily seen to be true if we regard $1 - P\{F\}$ as the probability of "not F" and realize that the minimum value for $P\{EF\}$ occurs when we put as much as possible of the probability for E into "not F." That is, we want maximum overlap between E and "not F." As an example of the use of this inequality, suppose we know that the probability of a signal source working is 0.8 and the probability of the signal detector working is 0.7, then the two will both work at least 50% of the time. Note that if the two are uncorrelated, they will both be working 56% of the time.

The next inequality is the **Markov inequality,** which states that if x is a non-negative random variable, then

$$P\{x \geq a\} \leq \frac{\mathbb{E}\{x\}}{a}. \tag{3.30}$$

This inequality gives a maximum value for the probability of the tail of a distribution. For x continuous, it is proven as follows:

$$\mathbb{E}\{x\} = \int_0^\infty xf(x)\,dx = \left[\int_0^a + \int_a^\infty\right] xf(x)\,dx \geq \int_a^\infty xf(x)\,dx$$

$$\mathbb{E}\{x\} \geq a \int_a^\infty f(x)\,dx = aP\{x \geq a\}.$$

Next apply Markov's inequality to the variable $x = (y - \mathbb{E}\{y\})^2$, where y has variance σ^2. Set $a = k^2$. Note that $\mathbb{E}\{x\} = \mathbb{E}\{(y - \mathbb{E}\{y\})^2 = \sigma^2\}$. We then have

$$P\{|y - \mathbb{E}\{y\}| \geq k\} \leq \frac{\sigma^2}{k^2}. \tag{3.31}$$

This is known as **Chebyshev's inequality**. Using this inequality, we see that the probability that a result differs by more than three standard deviations from the mean is less than 1/9 regardless of the distribution. However, if we know the distribution, the result can often be greatly improved. For example, for the normal distribution to be discussed in Chap. 6, the probability of greater than a three standard deviation from the mean is 0.0026.

In this chapter, we have examined results that do not depend on specific probability distributions. We have treated multiple scattering and examined results that flow from independence or (for $\overline{y\theta}$) dependence of the various scattering events. The fundamental root N law, which follows from the independence of the various events, is a very basic result that appears in many forms in many branches of physics and elsewhere. We will see it appear several more times in this course. Further examples of results that do not depend on specific probability distributions are given by propagation of errors and the related process of errors when changing variables. Finally, we examined some useful probability inequalities. It is now time for us to start looking at specific distributions, and we will start that process in the next chapter.

3.4   Worked Problems

WP3.1 Suppose we take n independent measurements of the same quantity. Suppose each measurement has mean $\bar{x}$ and variance σ_x^2.
We then average the result.

$$x_{AV} = (x_1 + x_2 + \cdots + x_n)/n. \tag{3.32}$$

Find the mean and variance of x_{AV}.

Answer: Let

$$G = (x_1 + \cdots + x_n)/n \equiv x_{AV}, \qquad \overline{G} = (\bar{x}_1 + \bar{x}_2 + \cdots + \bar{x}_n)/n = \bar{x},$$

$$\sigma_{x_{AV}}^2 = \sigma_G^2 = \sum_{i,j=1}^{n} \frac{\partial x_{AV}}{\partial x_i} \frac{\partial x_{AV}}{\partial x_j} \overline{(x_i - \bar{x}_i)(x_j - \bar{x}_j)}. \tag{3.33}$$

If the x_j are independent, $\overline{(x_i - \bar{x}_i)(x_j - \bar{x}_j)} = 0$ if $i \neq j$. (This is the trick from multiple scattering.)

$$\frac{\partial x_{AV}}{\partial x_i} = \frac{1}{n},$$

$$\sigma_{x_{AV}}^2 = \sum_{i=1}^{n} \left(\frac{1}{n}\right)\left(\frac{1}{n}\right)\sigma_x^2 = \frac{\sigma_x^2}{n}, \tag{3.34}$$

$$\sigma_{x_{AV}} = \frac{\sigma_x}{\sqrt{n}}.$$

The σ of the mean is $1/\sqrt{n}$ times the σ of the individual measurements. Independence is the essential ingredient for this very general relation.
As an extension to the above, it is amusing to look at

$$\sigma_{exp}^2 = \overline{(x_i - x_{AV})^2}.$$

Note:

$$x_{AV} \neq \bar{x}, \qquad x_{AV} = \frac{1}{n}\sum x_j = G. \tag{3.35}$$

$$\sigma_{exp}^2 = \underbrace{\overline{(x_i - \bar{x})^2}}_{\sigma_x^2} + \underbrace{\overline{(\bar{x} - x_{AV})^2}}_{\frac{\sigma_x^2}{n}} + \underbrace{\overline{2(x_i - \bar{x})(\bar{x} - x_{AV})}}_{\text{C.T.} = \text{crossterm}}.$$

Use independence. This implies that only the $j = i$ term contributes to the C.T.

$$C.T. = \overline{-2(x_i - \bar{x}) \sum_j \frac{(x_j - \bar{x})}{n}} = -\frac{2}{n}\sigma_x^2,$$

(3.36)

$$\sigma_{exp}^2 = \sigma_x^2\left[1 + \frac{1}{n} - \frac{2}{n}\right] = \sigma_x^2\left[1 - \frac{1}{n}\right] = \sigma_x^2\left[\frac{n-1}{n}\right].$$

Express $\sigma_{x_{AV}}^2$ in terms of σ_{exp}^2.

$$\sigma_{x_{AV}}^2 = \sigma_x^2/n, \qquad \sigma_{exp}^2 = \frac{n-1}{n}\sigma_x^2.$$

To estimate σ_x^2, use

$$\sigma_x^2 = \frac{n}{n-1}\sigma_{exp}^2 = \frac{n}{n-1}\frac{1}{n}\sum\overline{(x_i - x_{AV})^2}.$$

Thus,

$$\sigma_x^2 = \frac{1}{n-1}\sum\overline{(x_i - x_{AV})^2} \cong \frac{1}{n-1}\sum(x_i - x_{AV})^2.$$

(3.37)

$$\sigma_{x_{AV}}^2 = \frac{1}{n}\frac{n}{n-1}\sigma_{exp}^2 = \frac{\sigma_{exp}^2}{n-1} = \sum\frac{\overline{(x_i - x_{AV})^2}}{n(n-1)} \cong \frac{\sum(x_i - x_{AV})^2}{n(n-1)}.$$

(3.38)

This is similar to our previous relation for $\sigma_{x_{AV}}$, Eq. 3.35, except that $\sqrt{n}$ has been replaced by $\sqrt{n-1}$. Jargon: There are "n degrees of freedom," i.e., n independent variables. Using the experimental not the real mean uses up 1 degree of freedom (d.f.) because $\sum_i(x_i - x_{AV}) = 0$. That is why n goes to $n - 1$.

WP3.2 Suppose we make two independent measurements of the same quantity but they have different variances.

$$x_1 \ (\bar{x}, \ \sigma_1),$$
$$x_2 \ (\bar{x}, \ \sigma_2).$$

How should we combine them linearly to get a result with the smallest possible variance?

Answer:

$$x \equiv G = cx_1 + (1 - c)x_2.$$

Find c to minimize σ_G^2.
What is the $\overline{G}$ and σ_G^2? (Do you see a convenient way to extend this to n measurements?)

Example Two measurements for the ratio of neutral to charged current events for neutrinos interacting on nuclei are

$$0.27 \pm 0.02 \ CITF \ (Fermilab),$$
$$0.295 \pm 0.01 \ CDHS \ (CERN).$$

What would you quote for a combined result?

Answer: Let

$$x = cx_1 + dx_2.$$

We want the mean of x to be $\bar{x}$. This implies $d = 1 - c$.

$$x = cx_1 + (1 - c)x_2,$$
$$\sigma_x^2 = c^2\sigma_1^2 + (1 - c)^2\sigma_2^2,$$

since

$$\sigma_G^2 = \sum \left(\frac{\partial G}{\partial x_i}\right)^2 \sigma_i^2 \quad \text{if the } x_i \text{ are independent.}$$

At a minimum,

$$d\sigma_x^2/dc = 0 = 2c\sigma_1^2 - 2(1 - c)\sigma_2^2.$$

$$c = \frac{\sigma_2^2}{\sigma_1^2 + \sigma_2^2}, \qquad 1 - c = \frac{\sigma_1^2}{\sigma_1^2 + \sigma_2^2},$$

$$x = \frac{\sigma_2^2 x_1 + \sigma_1^2 x_2}{\sigma_1^2 + \sigma_2^2} = \frac{x_1/\sigma_1^2 + x_2/\sigma_2^2}{1/\sigma_1^2 + 1/\sigma_2^2}, \qquad (3.39)$$

$$\sigma_x^2 = \frac{1}{1/\sigma_1^2 + 1/\sigma_2^2}.$$

The generalization to n measurements is easy in this form:

$$x = \frac{\sum_i x_i/\sigma_i^2}{\sum_j 1/\sigma_j^2}, \qquad \frac{1}{\sigma_x^2} = \sum_i \frac{1}{\sigma_i^2}. \qquad (3.40)$$

The form is similar to resistors in parallel. The weight of each measurement is proportional to the inverse square of its error.

Example:

$$x = \left(\frac{0.27}{(0.02)^2} + \frac{0.295}{(0.01)^2}\right) \Big/ \left(\left(\frac{1}{0.02}\right)^2 + \left(\frac{1}{0.01}\right)^2\right) = 0.29,$$

$$\sigma = \sqrt{\frac{1}{(1/0.02)^2 + (1/0.01)^2}} = 0.009.$$

Another question we might ask is whether the two measurements are consistent. To do this, we need to look at the difference and the error in the difference of the two measurements. However, if we take the difference of the two measurements, the error is $\sigma_1^2 + \sigma_2^2$, just as it is with the sum.

$$x_1 - x_2 = (0.295 - 0.27) \pm \sqrt{(0.02)^2 + (0.01)^2} = 0.025 \pm 0.022.$$

3.5 Exercises

3.1 Suppose we obtain n independent results x_i from a distribution $F(x)$. Let x_{AV} be the mean of these n values, the sample mean. Define

$$\sigma_s^2 \equiv \frac{1}{n-1} \sum_{i=1}^{n} (x_i - x_{AV})^2. \qquad (3.41)$$

Show that the expectation value of σ_s^2 is σ^2.

Remark It can also be shown by similar (but lengthy calculations) that the variance of σ_s^2 is [to order of terms in $(1/n)^2$]

$$\mathrm{var}(\sigma_s^2) = \frac{1}{n-1} \left(\frac{n-2}{n-1} \mu_4 - \sigma^4 \right), \qquad (3.42)$$

where μ_2 and μ_4 are the second and fourth central moments of the distribution. This is a useful result for estimating the uncertainty of an error estimate.

3.2 Imagine that a charged particle is passing through matter undergoing multiple scattering. Neglect energy loss, i.e., let $\overline{\theta_y^2} = Kx$. We wish to estimate the direction of motion of the particle in the x-y plane by measuring its y position independently at two points, $x = 0$ and $x = L$. We estimate the particle direction by $\tan \theta = (y(L) - y(0))/L$. The y variance of each measurement is σ_{meas}^2. The x positions are known precisely. Assume all angles with respect to the x-axis are small. Show that an optimum length exists, i.e., a length at which the variance of $\tan \theta$ is a minimum. Find that length in terms of K and σ_{meas}^2.

3.3 Suppose one has measured a counting rate in a scintillation counter twice, obtaining $R_1 = m_1/T_1$ and $R_2 = m_2/T_2$, respectively, as results. m_1, m_2 are the number of counts obtained and T_1, T_2 are the times used for the measurements. Suppose the errors in T_1, T_2 are negligible and that $\sigma_{m_1}^2 \approx m_1, \sigma_{m_2}^2 \approx m_2$. Show that if these measurements are approximately consistent, i.e., $m_1/T_1 \approx m_2/T_2$, then the best way to combine the data is as one big observation: $R = (m_1 + m_2)/(T_1 + T_2)$. Find the variance.

3.4 Suppose that we are measuring counting rates with a scintillator and that we are measuring a signal and a background. First we measure a background plus a signal, then a background only, obtaining $R_1 = m_1/T_1$ and $R_2 = m_2/T_2$, respectively, as results. m_1, m_2 are the number of counts obtained and T_1, T_2 are the times used for the measurements. Suppose the errors in T_1, T_2 are negligible and that $\sigma_{m_1}^2 \approx m_1$, $\sigma_{m_2}^2 \approx m_2$. For a given total time of observation $T = T_1 + T_2$, how should we divide our time between T_1 and T_2 to get the most accurate value for $R(= R_1 - R_2)$?

3.5 Atomic physicists have measured parity violation effects in a series of delicate experiments. In one kind of experiment, they prepare two samples identically except for reversed external fields and look at the intensity of fluorescence from each when illuminated by a laser pulse. Suppose they measure about 10^6 photons per laser shot and look for a 10^{-6} effect in the difference/sum, i.e., the value of $(N_1 - N_2)/(N_1 + N_2)$. If the variance of N is about N, and ignoring systematic effects, how many laser pulses do they need to have a non-zero effect statistically valid to about three standard deviations?

3.6 Two measurements for κ, the compressibility of liquid He, gave $3.87 \pm 0.04 \times 10^{-12}$ and $3.95 \pm 0.05 \times 10^{-12}$ cm^2/dyn. From these two experiments, find the best estimate and its error for the value of κ. Are these two experiments consistent with each other? The standard value for κ is 3.88×10^{-12} cm^2/dyn. Is the combined result consistent with this?

References

Einstein A (1905) Die von der molekularkinetischen theorie der wärme geforderte bewegung von in ruhenden flüssigkeit suspendierten teilchen. Ann Phys 17:549–560
Pearson K (1905) Nature 72

Chapter 4
Discrete Distributions and Combinatorials

Abstract Most of the applications of probability theory require that we understand the properties of a few specific distributions that appear in a number of applications and that we further understand how to derive distributions if we need new ones. The concept of combinatorials is central to this task. A few special discrete probability functions will be defined. We will also discuss Stirling's approximation for the factorial function.

4.1 Combinatorials

We start by considering some combinatorial properties. Consider n objects, all of which are different (i.e., distinguishable). Imagine they are all placed in a box and we take out r of them one at a time. We do this without replacement, that is, when we take an object out we leave it out; we do not put it back into the box. How many different ways, n_r, are there of doing this?

$$n_r = n(n-1)(n-2)\cdots(n-r+1) = \frac{n!}{(n-r)!}. \qquad (4.1)$$

This is easily seen as the first time you have a choice of n objects. For the second time, one has already been removed, so you have a choice of $n-1$. The third trial gives $n-2$, and so on. Here

$$k! \equiv k(k-1)(k-2)\cdots 1, \qquad 0! \equiv 1. \qquad (4.2)$$

Next we ask how many different sets of r objects can be picked from the above n? Note this is *not* the same question as the previous one. For example, if $r = 2$, we could pick object 1 then object 2, or object 2 then object 1. These are two different ways of picking the objects, but both lead to the same set of objects picked.

© Springer Nature Switzerland AG 2020
B. P. Roe, *Probability and Statistics in the Physical Sciences*,
Undergraduate Texts in Physics,
https://doi.org/10.1007/978-3-030-53694-7_4

The number of different sets of r objects from n is called $\binom{n}{r}$ or the "binomial coefficient." It equals n_r/(the number of ways of picking the same r objects). This latter quantity is $r!$; for the first try, you pick any of the r objects; for the second try, any of the $r-1$ remaining; etc. Therefore, the binomial coefficient becomes

$$\binom{n}{r} = \frac{n!}{(n-r)!r!}. \tag{4.3}$$

Why is it called the binomial coefficient? Consider $(x+y)^n$. Look at the term $x^{n-r}y^r$ and ask what is the numerical coefficient in front of that term. For definiteness, take $n=5$, $r=3$. We are then considering $(x+y)(x+y)(x+y)(x+y)(x+y)$. A typical term with $r=3$ is $xyyxy$, i.e., x from the first and fourth terms, y from the second, third, and fifth terms. The numeral coefficient in front of x^2y^3 is the number of sets of three y's picked from five terms. The logic is just the same as above. The first y can be picked from any of the five terms, the second from any of the remaining four, etc. You must divide by $r(\equiv 3)!$ because this gives the number of ways, using the above prescription, of getting the same sequence, e.g., $xyyxy$. For example, you could choose 2, 3, 5 for the y positions or 5, 2, 3. Both give the same sequence.

$$(x+y)^n = \sum_{r=0}^{n} \binom{n}{r} x^{n-r} y^r. \tag{4.4}$$

We turn now to a new problem. Suppose we have n cells and r distinguishable objects. How many arrangements of objects in the cells are possible if each cell can have any number of objects? The answer is n^r since any of the r objects can be in any of the n cells. Suppose each arrangement is equally probable. Let P_k be the probability that a given cell (i) has k objects.

$$P_k = \frac{(\text{number of arrangements with } k \text{ objects in cell } i)}{(\text{total number of arrangements})}. \tag{4.5}$$

The numerator is the number of sets of k objects chosen from r distinguishable objects times the number of ways of arranging the remaining $r-k$ objects in the remaining $n-1$ cells. The denominator is n^r.

$$P_k = \binom{r}{k} \frac{(n-1)^{r-k}}{n^r}. \tag{4.6}$$

Let us consider some examples of this type of computation using bridge hands of playing cards. Decks of playing cards have 52 cards, divided into four 13 card suits, spades, hearts, diamonds, and clubs. The cards in each suit are denoted 1–10, jack, queen, and king. The game of bridge has four players, and the 52 cards in the deck are randomly distributed into four hands of 13 cards each, one for each player

The number of possible hands is $\binom{52}{13}$.

The probability of 5 diamonds, 5 spades, 2 hearts, and 1 club is

$$\binom{13}{5}\binom{13}{5}\binom{13}{2}\binom{13}{1}\Big/\binom{52}{13}. \tag{4.7}$$

The probability of 5 spades and 8 non-spades is

$$\binom{13}{5}\binom{39}{8}\Big/\binom{52}{13}. \tag{4.8}$$

The probability of A, 2, 3, 4 ..., J, Q, K with no suit restriction is

$$\frac{4^{13}}{\binom{52}{13}}. \tag{4.9}$$

4.2   Stirling's Approximation for the Factorial Function

The probabilities considered in the previous section all involve factorials with high numbers. There is a very convenient approximation for $n!$ called Stirling's approximation. It can be shown that

$$n! \cong \left(\frac{n}{e}\right)^n \sqrt{2\pi n}\, e^{1/(12n)}. \tag{4.10}$$

The first term $(n/e)^n$ is called the zeroth approximation, the first two terms $(n/e)^n\sqrt{2\pi n}$ the first approximation, and all three terms the second approximation.

The first approximation is the standard one used. It is clear that even for small n we get a good approximation and that for $\log(n!)$ (we use base e entirely for logarithms here) if n is greater than about 100, the zeroth approximation is accurate to better than 1%. See Table 4.1.

In this chapter, we have looked at problems of combinatorials and derived a few probability distributions for very simple situations. In the next chapter, we shall look at some of the standard one dimensional discrete probability distributions.

Table 4.1 Approximations for factorials

n	$n!$	First approx.	Error
1	1	0.922	8
2	2	1.919	4
3	6	5.84	2.5
4	24	23.51	2.1
5	120	118.02	2.0
10	3.6288×10^6	3.599×10^6	0.8
100	9.3326×10^{157}	9.3249×10^{157}	0.08

n	Second approx.	% Error	$n(\log n) - n$	$1/2 \log(2\pi n)$
1	1.0023	0.2	-1	0.919
2	2.0007	0.04	-0.61	1.27
3	6.0005	0.01	0.3	1.47
4	24.001	0.004	1.545	1.61
5	124.003	0.002	3.04	1.72
10	3.62881×10^6	–	13	2.1
100	9.3326×10^{157}	–	362	3.2

4.3   Worked Problems

WP4.1(a.) Find the probability expression for having 4 kings in a bridge hand.

WP4.1(b.) Find the probability expression for having 4 kings or 4 aces (or both) in a bridge hand.

Answer

4.1a.

$$P_{4 \text{ kings}} = \frac{\binom{48}{9}}{\binom{52}{13}},$$

$$P_{4 \text{ kings}} = \frac{\frac{48!39!9!}{\frac{52!}{39!13!}}} = \frac{48!}{52!} \frac{13!}{9!} = \frac{13 \times 12 \times 11 \times 10}{52 \times 51 \times 50 \times 49} = 0.00264105.$$

4.1b.

$$P_{\substack{4 \text{ kings} \\ \text{and/or 4 aces}}} = \frac{2 * \binom{48}{9}}{\binom{52}{13}} - \frac{\binom{44}{5}}{\binom{52}{13}}.$$

The last term subtracts the double counted part.

WP4.1(c.) Using Stirling' approximation evaluate the expression obtained in 4.1a. (Hint: It is easiest to calculate log P first.)

Answer

Using the Stirling approximation:

$$\log P_{4 \text{ kings}} = \log 48! + \log 13! - \log 52! - \log 9!$$

$$= 48 \log \frac{48}{e} + 13 \log \frac{13}{e} - 52 \log \frac{52}{e} - 9 \log \frac{9}{e}$$

$$+ \frac{1}{2}[\log 48 + \log 13 - \log 52 - \log 9]$$

(Note that the $\sqrt{2\pi}$'s cancel.)

$$= 48.5 \log 48 + 13.5 \log 13 - 52.5 \log 52$$

$$- 9.5 \log 9 - 48 - 13 + 52 + 9$$

$$= -5.93386,$$

$$P = 0.00264823.$$

4.4 Exercises

4.1 Consider the probability of players having complete suits, A, ..., K in bridge hands in a given round.

(a) Find the probability that one particular player has a complete suit.
(b) Find the probability that two specific players of the four bridge players have complete suits
(c) Find the probability that three specific players of the four have complete suits. (If three players have a complete suit, the fourth does also.)
(d) Find the probability that at least one of the four players has a complete suit.

4.2 Assume that we start with r red and b black balls in an urn. We randomly draw balls out one at a time without looking into the urn.

(a) We draw them out and place them to one side of the urn (no replacement). Find the probability of drawing n_1 black balls in N drawings. Assume that $n_1 < b$; $N - n_1 < r$.
(b) We draw them out and then replace the balls each time mixing them up in the urn. (with replacement). Find the probability of drawing n_1 black balls in N drawings.

Chapter 5
Specific Discrete Distributions

Abstract In this chapter, we discuss several discrete distributions including the binomial distribution, the Poisson distribution, the multinomial distribution, the hypergeometric distribution, and the negative binomial distribution, the latter distribution appearing in the exercises.

5.1 Binomial Distribution

We have now developed the tools to derive some of the standard one-dimensional discrete probability distributions. We start by examining the binomial distribution and its limit, the Poisson distribution, which are two of the most common distributions we run across in applications.

We will first define Bernoulli trials. These are repeated independent trials, each of which has two possible outcomes. The probability of each outcome remains fixed throughout the trials. Examples of this are coin tossing, decay of k^+ into either $\mu^+ + \nu$ or another mode (i.e., these are the two possible outcomes). The results of each of these can be described as success (S) or failure (F).

Let p equal the probability of success and $q = 1 - p$ equal the probability of failure.

What is the probability of r successes in n trials? For example, consider the probability of three successes in five trials. FSSFS is one pattern leading to the result; SSSFF is another. They all have $p^3 q^2$ and we want the numerical coefficient of r successes in n trials. The argument proceeds exactly as in the discussion of $(x + y)^n$.

$$P_r = \binom{n}{r} p^r q^{n-r} = \frac{n!}{(n-r)!r!} p^r q^{n-r}. \tag{5.1}$$

© Springer Nature Switzerland AG 2020
B. P. Roe, *Probability and Statistics in the Physical Sciences*,
Undergraduate Texts in Physics,
https://doi.org/10.1007/978-3-030-53694-7_5

Not surprisingly, this is called the binomial distribution.

$$\sum_{r=0}^{n} P_r = (p+q)^n = 1. \tag{5.2}$$

Hence, we see that the distribution is appropriately normalized as, of course, it must be from the definition. Let us find the mean (m) and the standard deviation (σ).

Let t be a random variable that is defined as being 1 if the trial is a success and 0 if the trial is a failure.

Let

$$r \equiv \text{number of successes} = \sum_{i=1}^{n} t_i, \tag{5.3}$$

and

$$\bar{r} = n\bar{t}, \tag{5.4}$$
$$\bar{t} = 1*p + 0*q = p.$$

Thus, $\mathbb{E} = m = \bar{r} = np$.

$$\sigma^2 = \overline{(r-m)^2} = \overline{r^2} - (\bar{r})^2 = \overline{r^2} - (np)^2,$$
$$r^2 = \sum_{i,j} t_i t_j = \sum_{i \neq j} t_i t_j + \sum t_i^2. \tag{5.5}$$

Note that $t_i^2 = t_i$.

$$\sum \overline{t_i^2} = \sum \overline{t_i} = np. \tag{5.6}$$

i and j are independent for $i \neq j$.

$$\sum_{i \neq j} \overline{t_i t_j} = p^2 \sum_{i \neq j} 1 = p^2(n^2 - n),$$
$$\sigma^2 = p^2(n^2 - n) + np - (np)^2 = np(1-p).$$
$$\sigma^2 = npq. \tag{5.7}$$

Thus, $\dfrac{m}{\sigma} = \dfrac{np}{\sqrt{npq}} = \sqrt{n}\sqrt{\dfrac{p}{q}}.$

Note that

As n gets big, the peak gets sharper.

Let us take an example from statistical mechanics. Suppose we have a box containing a gas of non-interacting molecules. We count a success if a molecule is found in the left half of the box ($p = \frac{1}{2}$) and a failure if it is on the right side. A snapshot at a given instant has a trial for each molecule. If there are 10^{24} molecules, then $m/\sigma = 10^{12}$. We find the same number of molecules in each half of the box to about one part in 10^{12}. We started with a random distribution and have obtained startling regularity. This kind of result—statistical order from individual chaos—is a prime pillar upon which all statistical mechanics is built. Reif (1967) gives the following numerical example with the above model. Suppose there are only 80 particles and we take 10^6 snapshots a second. We will find a picture with all the molecules on one side only about one time in the life of the universe, i.e., about 10^{10} years.

The binomial distribution also occurs in a problem in which we have n components and the probability of failure of each is p. The probability of r component failures is then given by a binomial distribution.

The next distribution appears as the limit of the binomial distribution for the case

$$p \ll 1,$$
$$n \gg 1,$$
$$r \ll n,$$
$$pn = \lambda, \text{ a finite constant.} \tag{5.8}$$

An example of this limit is the number of decays per second from a radioactive element with a 1-year mean life. Here $n \sim 10^{23}$, $p = 1$ second/(number of seconds in a mean life) $\sim 0.3 \times 10^{-7}$, r is the number of decays in a 1 second trial and is about the size of pn, i.e., $\sim 3 \times 10^{15} \ll n$.

$$
\begin{aligned}
P_r &= \frac{n!}{r!(n-r)!} p^r q^{n-r} \\[2mm]
&= \frac{n!}{r!(n-r)!} \left(\frac{\lambda}{n}\right)^r \left(1 - \frac{\lambda}{n}\right)^{n-r} \\[2mm]
&= \underbrace{\frac{\lambda^r}{r!}}_{} \underbrace{\frac{n!}{(n-r)!n^r}}_{T1} \underbrace{\left(1 - \frac{\lambda}{n}\right)^{-r}}_{T2} \underbrace{\left(1 - \frac{\lambda}{n}\right)^{n}}_{T3}.
\end{aligned} \tag{5.9}
$$

Consider $T1$, $T2$, and $T3$ as $n \to \infty$, $\lambda (= pn)$ fixed, r fixed.

$$T1 = \frac{n}{n} \frac{n-1}{n} \frac{n-2}{n} \cdots \frac{n-r+1}{n} \to 1 \quad (r \text{ fixed}),$$

$$T2 = \left(1 - \frac{\lambda}{n}\right)^{-r} \to 1,$$

$$T3 = \left(1 - \frac{\lambda}{n}\right)^{n} \to e^{-\lambda}. \quad \text{(This is the limit used in the definition of } e.\text{)}$$

$$\text{Thus,} \qquad P_r \to \frac{\lambda^r}{r!}e^{-\lambda}. \tag{5.10}$$

This is known as the Poisson distribution.

$$\sum_{r=0}^{\infty} P_r = e^{-\lambda} \sum_{r=0}^{\infty} \frac{\lambda^r}{r!} = e^{-\lambda}e^{\lambda} = 1 \qquad \text{as it should.} \tag{5.11}$$

$$\mathbb{E} = m = \bar{r} = np = \lambda,$$

$$\sigma^2 = npq \to \lambda \qquad \text{as } q \to 1.$$

The Poisson distribution appeared as the limiting case of the binomial distribution. It also appears as an exact distribution, rather than a limiting distribution, for a continuous case counting problem when the Poisson postulate holds. The Poisson postulate states that whatever the number of counts that have occurred in the time interval from 0 to t, the conditional probability of a count in the time interval from t to $t + \Delta t$ is $\mu \Delta t + 0(\Delta t^2)$. μ is taken to be a fixed constant. $0(\Delta t^2)$ means terms of higher order in Δt.

$$P_r(t + \Delta t) = P_r(t)P_0(\Delta t) + P_{r-1}(t)P_1(\Delta t) + 0(\Delta t^2)$$

$$= P_r(t)(1 - \mu \Delta t) + P_{r-1}(t)\mu \Delta t + 0(\Delta t^2).$$

Thus,
$$\frac{dP_r}{dt} = -\mu P_r + \mu P_{r-1}, \tag{5.12}$$

where for

$$r = 0, \ P_{r-1} \text{ is defined to be } 0.$$

The solution for $r = 0$ is easily seen to be $P_0 = e^{-\mu t}$ and the Poisson form with $\lambda = \mu t$ can be shown to follow by induction.

For λ large and $r - \lambda$ not too large, the Poisson distribution approaches another distribution called the normal or Gaussian distribution with the same m and σ. The normal distribution and other continuous distributions are the subject of the next chapter.

Suppose we are evaporating a thin film onto a substrate. Imagine that the gas molecules have a probability 1 of sticking to the substrate if they hit it. Consider the probability that in a small area (about the size of one molecule), we have a thickness of n molecules. Here we have a great many molecules in the gas, each having only a very small probability of landing in our chosen area. The probabilities are independent for each molecule and the conditions for a Poisson distribution are satisfied. The distribution of the thickness in terms of the number of molecules is Poisson.

The Poisson distribution also occurs in diffusion. In a gas, the probability of collision in dt can be taken to be dt/τ. If we make the stochastic assumption that this probability holds regardless of how long it has been since the last collision, then the distribution of the number of collisions in time T will be a Poisson distribution with mean value T/τ.

In this chapter, we have looked at the properties of the binomial distribution and the Poisson distribution, two of the most common discrete distributions in practice. In the worked problems, we introduce the multinomial distribution and the hypergeometric distribution. In the homework exercises, we introduce the negative binomial distribution (related to the binomial distribution).

5.2 Worked Problems

WP5.1(a.) Suppose there are n objects of k distinguishable kinds, n_1 of kind 1, n_i of kind i. How many distinguishable orderings are there (multinomial coefficients)?

Answer

$$\frac{n!}{n_1! \, n_2! \cdots n_k!} = \text{the number of orderings}$$

$$= \frac{\text{the total number of orderings}}{(\text{permutation of kind 1}) \cdots},$$

$$(x_1 + x_2 + \cdots + x_k)^n = \sum \frac{n!}{n_1! n_2! \cdots n_k!} x_1^{n_1} x_2^{n_2} \cdots x_k^{n_k}.$$

WP5.1(b.) Suppose we now randomly pick r of those n objects with replacement. What is the probability of having r_1 of kind $1, \ldots, r_i$ of kind i, etc., where $\sum_{i=1} r_i = r$ (multinomial distribution)?

Answer

Consider the arrangements of the r objects picked. The probability of each arrangement is $p_1^{r_1} p_2^{r_2} \cdots p_k^{r_k}$ where $p_i = n_i/n$. The probability of the arrangement times the number of arrangements = the total probability.

$$P = \left(\frac{r!}{r_1! r_2! r_3! \ldots r_k!} \right) p_1^{r_1} p_2^{r_2} \cdots p_k^{r_k}. \tag{5.13}$$

WP5.1(c.) Try part b if the objects are picked without replacement (hypergeometric distribution).

Answer

This is similar to our bridge game problems. The binomial coefficient $\binom{n_i}{r_i}$ is the number of ways of picking r_i objects from n_i objects.

$$P = \frac{\binom{n_1}{r_1}\binom{n_2}{r_2}\cdots\binom{n_k}{r_k}}{\binom{n}{r}}. \tag{5.14}$$

This is useful in quality control. We have n objects and n_1 defectives. How big a sample must be taken to have r_1 defectives?

WP5.2 Suppose we have a radioactive source with a long lifetime emitting particles, which are detected by a scintillation counter and then counted in a scaling circuit. The number of counts/second detected has a Poisson distribution.

The scalar counts to a certain number s and then gives an output pulse (which might go to the next stage of a scalar). What is the distribution of spacing in time between these output pulses of the scalar?

Answer   Let:

$\overline{n_0}$ = mean output rate,

$\overline{n_I} = s n_0$ = the mean input rate (counts/sec),

t = the time since the last output pulse,

$f_s(t)$ = density function for the time until the next output pulse, i.e., the pacing between pulses.

The probability for n input counts into the scalar at time t is

$$P_n = \frac{\lambda^n e^{-\lambda}}{n!}, \qquad \text{where } \lambda = \overline{n_I} t = s\overline{n_0} t. \tag{5.15}$$

$f_s(t)\,dt$ is (the probability of $s-1$ counts in 0 to t) times (the probability of 1 count in dt).

$$f_s(t)dt = \frac{\lambda^{s-1} e^{-\lambda}}{(s-1)!} * \overline{n_I}\,dt,$$

$$f_s(t)\,dt = \overline{n_I}\frac{(\overline{n_I}t)^{s-1} e^{-\overline{n_I}t}}{(s-1)!}\,dt. \tag{5.16}$$

The distribution with density function

$$f(y) = \frac{\mu e^{-\mu y}(\mu y)^{\alpha-1}}{(\alpha-1)!}, \qquad y \geq 0 \tag{5.17}$$

is known as the gamma distribution and is illustrated in Fig. 5.1 for $s = 1, 2, 3, 5$. The mean of the gamma distribution is α/μ, and the variance is α/μ^2. For $\alpha = 1 (= s)$, the gamma distribution becomes $f = \mu e^{-\mu x}$, which is known as the exponential distribution. This distribution will be discussed in Sect. 6.9.1. For $\mu = 1/2$; $\alpha = n/2$, the gamma distribution becomes the χ^2 distribution, which we shall discuss in the next chapter.

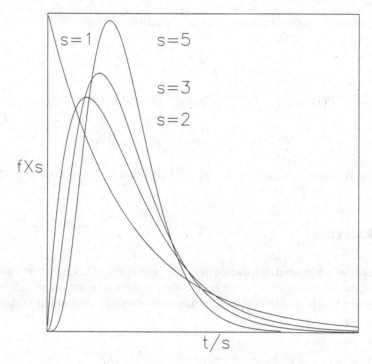

Fig. 5.1 A plot of the gamma distribution for different s values for $\overline{n_I} = 1$. The abscissa is t/s and the ordinate is $f \times s$

With no scaling, the most probable interval is 0. In terms of $x = \overline{n}_0 t$, the peak is at $x_{max} = 1 - 1/s$. For large s, this is a sharp peak (almost Poisson). Therefore, the higher stages of a scaling circuit see regularly spaced pulses. Most of the special electronics design need to go into only the first few stages. For large s, we get an accurate clock! We get great regularity and accuracy even though we start with a random process.

WP5.3 We have a crystal that has two types of impurities, one of which will absorb a particular frequency photon without freeing an electron and the other which will free an electron when absorbing the photon. There are the same number of impurities of each kind, but the cross section for absorption in the first kind of impurity is 99 times larger than the cross section for absorption in the second kind of impurity. The crystal is thick enough that all the photons are absorbed. Suppose 200 photons are incident. What is the probability of at least three electrons being liberated?

Answer

$$P(\geq 3) = 1 - P(0) - P(1) - P(2).$$

The probability is distributed in a binomial distribution with $p = 0.01$ and $n = 200$.

$$P(0) = (1 - p)^n = (0.99)^{200} = 0.134,$$

$$P(1) = np(1 - p)^{n-1} = 200 \times 0.01 \times (0.99)^{199} = 0.271,$$

$$P(2) = \frac{n(n - 1)}{2} p^2 (1 - p)^{n-2} = 0.267.$$

$$P(\geq 3) = 1 - P(0) - P(1) - P(2) = 1 - 0.134 - 0.271 - 0.267 = 0.328.$$

5.3   Exercises

5.1 The negative binomial distribution gives the probability of having the mth success occur on the rth Bernoulli trial, where p is the probability of the occurrence of the success of a single trial. Show that the negative binomial distribution probability function is given by

$$P(r, m) = \frac{(r - 1)!}{(m - 1)!(r - m)!} p^m (1 - p)^{r-m}. \tag{5.18}$$

The mean of this distribution is m/p and the variance is $m(1 - p)/p^2$. Hint: Consider the situation after trial $r - 1$.

5.2 A nuclear physics experiment is running. It requires 50 independent counters sensitive to ionizing radiation. The counters are checked between data runs and are observed to have a 1% chance of failure between checks (i.e., $p = 0.01$ for failure). What is the probability that all the counters were operative throughout a data run? Suppose that a data run is salvageable if one counter fails, but is not useful if two or more counters fail. What is the probability that the data run is spoiled by counter failure?

5.3 We consider collisions of a high-energy proton with a proton at rest. Imagine the resultant products consist of the two protons and some number of pions. Suppose we imagine the number of pions is even and equal to $2N$, where N is the number of pion pairs. As a crude model, we imagine the probability of N pairs is Poisson distributed with mean $= \lambda$. (In practice, λ is small enough that energy conservation and phase space do not limit the number of pairs in the region of appreciable probabilities.) Suppose each of the N pion pairs has a probability p of being charged ($\pi^+ \pi^-$) and $q = 1 - p$ of being neutral ($\pi^\circ \pi^\circ$). Show that the probability of n charged pairs of pions in the collision is Poisson distributed with mean value $= p\lambda$.

5.4 Suppose we are measuring second harmonic generation when pulsing a laser on a solid. However, the detector we have available is only a threshold detector and the time resolution is such that it can only tell us whether or not there were any second harmonic photons on that pulse. Assume the generated second harmonic photons have a Poisson distribution. Show that by using a number of pulses and measuring the probability of count(s) versus no count, we can measure the average number of second harmonic photons per pulse. (In practice, this method was used successfully over a range of six orders of magnitude of intensity of the final photons.)

5.5 This problem has arisen in evaluating random backgrounds in an experiment involving many counting channels. Suppose one has three channels and each of the three channels is excited randomly and independently. Suppose at any instant the probability of channel i being on is p_i. Imagine "on" means a light on some instrument panel is on. To simplify the problem, take $p_1 = p_2 = p_3 = p$. At random intervals, pictures are taken of the instrument panel with the lights on it. In a given picture, what are the probabilities of 0, 1, 2, 3 channels being on? N total pictures are taken. What is the joint probability that n_0 pictures have 0 channels on, n_1 have 1 channel on, n_2 have 2 channels on, and n_3 have 3 channels on, where we must have $n_0 + n_1 + n_2 + n_3 = N$?

Reference

Reif F (1967) Statistical physics. Berkeley physics course, vol 5. McGraw-Hill, New York

Chapter 6
The Normal (or Gaussian) Distribution and Other Continuous Distributions

Abstract In this chapter, a number of continuous distributions will be discussed. The extremely important normal distribution will be the first to be discussed. After that a large number of other continuous distributions will be treated including the chi-squared distribution, the F distribution, Student's distribution, the Cauchy (Breit–Wigner) distribution, and several others.

6.1 The Normal Distribution

By far, the most important of the continuous distributions is the normal or Gaussian distribution that appears almost everywhere in probability theory, partially because of the central limit theorem, which will be discussed in Chap. 11. In addition, we will introduce a few other distributions of considerable use. The chi-square distribution, in particular, as we will see later, plays an important role when we consider the statistical problem of estimating parameters from data.

The normal distribution is a continuous distribution that appears very frequently. As with the Poisson distribution, it often appears as a limiting distribution of both discrete and continuous distributions. In fact, we shall find that under very general conditions a great many distributions approach this one in an appropriate limit.

We shall give here a simple discussion showing that many distributions approach the normal one as a limit, which indicates something of the generality of this phenomenon. We shall return to this point and quote a more complete theorem in Chap. 11.

Consider P_n, the probability in a discrete distribution with a very large number of trials N for a number of successes n. Here P_n is not necessarily the binomial distribution, but any "reasonably behaved singly peaked distribution." We shall leave the description vague for the moment.

Let P_x be a reasonably smooth function taking on the values of P_n at $x = n$. We expand P_x in a Taylor series near the distribution maximum. However, as we have seen for the binomial distribution, the function often varies quickly near the maximum for large N and we will expand $\log P$ as it varies more slowly, allowing the series to converge better.

© Springer Nature Switzerland AG 2020
B. P. Roe, *Probability and Statistics in the Physical Sciences*,
Undergraduate Texts in Physics,
https://doi.org/10.1007/978-3-030-53694-7_6

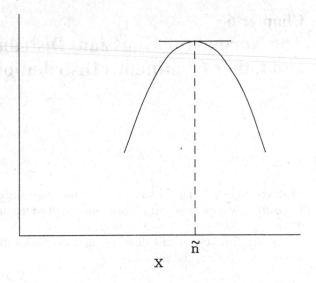

Fig. 6.1 A typical peaked distribution P near the maximum $\tilde{n}$

$$\log P_x = \log P_{\tilde{n}} + \frac{B_1\eta}{1!} + \frac{B_2\eta^2}{2!} + \cdots,$$

where

$$\tilde{n} = \text{value of } x \text{ at maximum of } P,$$

$$\eta = n - \tilde{n},$$

$$B_k = \left.\frac{d^k \log P}{dx^k}\right|_{\tilde{n}}.$$

(6.1)

It is easily seen that $B_1 = 0$ and $B_2 < 0$ since $P_{\tilde{n}}$ is a maximum (see Fig. 6.1). The crucial assumption is that the distribution is sharp enough that $\log P_x$ can be expanded ignoring terms higher than second order. We now ignore terms higher than η^2. Then

$$P_n \cong P_{\tilde{n}}\, e^{-|B_2|\eta^2/2}.$$

(6.2)

We now wish to go over into a continuous distribution. We assume $P_n \cong P_{n+1}$ near the maximum. (This does not contradict the existence of a sharp peak. $\tilde{n}$ is huge.)

We let $P_n \Delta n \sim f\, dx$, where f and dx are defined in terms of a step length:

$$x = n\ell,$$

$$\ell = \text{step length (e.g., the multiple scattering problem)},$$

$$f = \text{probability density function of } x = \frac{P_n}{\ell},$$

$$dx = \ell \times \Delta n$$

(6.3)

Thus,

$$f = Ce^{-(n-\tilde{n})^2|B_2|/2} = Ce^{-(x-\tilde{x})^2/2\sigma^2}, \tag{6.4}$$

where

$$C = \text{a constant,}$$
$$\tilde{x} = \tilde{n}\ell,$$
$$\sigma^2 = \ell^2/|B_2|,$$

C should be obtained by requiring the integral of f over all allowed x to be unity. However, since the peak tends to be fairly sharp, the integral over all allowed x can be replaced by the integral over all x since the wings of the distribution are small.

$$1 = \int_{-\infty}^{\infty} f\, dx = \int_{-\infty}^{\infty} Ce^{-(x-\tilde{x})^2/2\sigma^2}\, dx = \sqrt{2\sigma^2}C \int_{-\infty}^{\infty} e^{-y^2}\, dy.$$

We have seen previously that this leads to

$$f(x) = \frac{1}{\sqrt{2\pi\sigma^2}} e^{-(x-\tilde{x})^2/2\sigma^2}. \tag{6.5}$$

The distribution *defined* by the above density function is called the normal or Gaussian distribution. What we have shown above is that many distributions approach the normal distribution in an appropriate limit. The major assumption was that, near the peak, log P_x could be expanded ignoring terms higher than second order. The mean and variance are easily obtained:

$$\mathbb{E} = mean = m = \tilde{x}$$
$$\text{This follows since } f \text{ is symmetric in } (x - \tilde{x}). \tag{6.6}$$
$$\text{var}(x) = \sigma^2.$$

The mean is the most efficient estimator of $\mathbb{E}$ and is an unbiased estimator. The variance relation will be derived in Worked Problem 6.1.

For the normal distribution the probabilities of a result being more than k standard deviations from the mean are 0.3174 for $k = 1$, 0.0456 for $k = 2$, and 0.0026 for $k = 3$. The probability of being in the range $(m \pm \sigma) = .6827$. $P(x$ in range $m \pm 0.6745\sigma) = 0.5$, $\mathbb{E}[|x - m|] = (2/\pi)\sigma) = 0.7979\sigma$. The half-width at half-maximum $= \sqrt{2}(\ln 2)\sigma = 1.177\sigma$.

The kurtosis (the fourth central moment/σ^4) $- 3$ is zero for a Gaussian distribution, positive for a leptokurtic distribution with longer tails, and negative for a platykurtic distribution with tails that die off more quickly than those of a Gaussian.

In Figs. 6.2, 6.3, and 6.4, the binomial and normal distribution are compared for increasing numbers of trials.

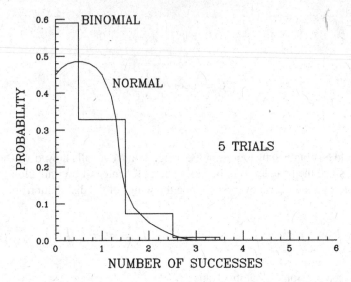

Fig. 6.2 A comparison of the binomial and normal density functions for five trials with probability of success 0.1 per trial

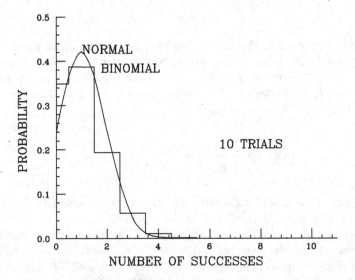

Fig. 6.3 A comparison of the binomial and normal density functions for 10 trials with probability of success 0.1 per trial

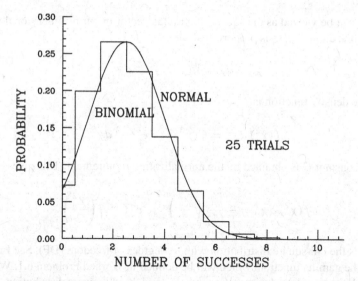

Fig. 6.4 A comparison of the binomial and normal density functions for 25 trials with probability of success 0.1 per trial

We indicated previously that when evaporating thin films onto a substrate, the thickness of the film had a Poisson distribution in terms of the number of molecules of thickness. We see now that as the film gets thicker, the distribution approaches a normal distribution, and for mean thicknesses N, where N is large, the thickness thus becomes quite uniform.

6.2 The χ^2 Distribution

We often have several measurements of a physical quantity, sometimes at each of several values of a parameter. We are frequently interested in some measure of the overall deviation of the measurements from the expected value of the physical quantity. The chi-squared distribution is a frequently used measure of this deviation. We will use this distribution extensively when we are comparing an experimental histogram with a theory.

Let $\chi^2 = z_1^2 + z_2^2 + \cdots + z_n^2$, where z_i is a number randomly chosen from a normal distribution with mean = 0 and variance = 1. If the variance is known to be not 1 but σ^2, we can replace the variables by z_i^2/σ_i^2.

We imagine the z_i to be mutually independent.

The density function for χ^2 is obtained from the normal density function:

$$f(z)\, d^n z \propto e^{-\chi^2/2}\, d^n z. \tag{6.7}$$

Here χ^2 can be viewed as the square of a radius vector in our n-dimensional space. The volume element $d^n z$ is proportional to

$$\chi^{n-1} d\chi = \tfrac{1}{2}(\chi^2)^{(n-2)/2} d\chi^2.$$

Thus, the density function is

$$f(\chi^2) d\chi^2 = C(\chi^2)^{n/2-1} e^{-\chi^2/2} d\chi^2,$$

and the constant C is obtained by the normalization requirement. This gives

$$f(\chi^2) d\chi^2 = \frac{1}{\Gamma\left(\frac{n}{2}\right)} \left(\frac{\chi^2}{2}\right)^{n/2-1} e^{-\chi^2/2} d\left(\frac{\chi^2}{2}\right). \qquad (6.8)$$

This is the chi-square distribution with n degrees of freedom (DF). See Fig. 6.5. $\Gamma(z)$ is the gamma function, which will be defined in Worked Problem 6.1. We note that the derivation started with $e^{-\chi^2/2}$ and obtained the chi-square distribution, which has much longer tails than implied by the exponential. This dependence came from the integration over n-dimensional space. Each point at a given radius away from 0 has only a small probability, but there are lots of points. It is necessary to keep this in mind in curve fitting applications.

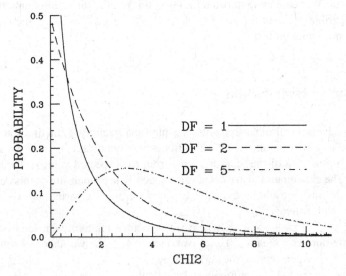

Fig. 6.5 The chi-square distribution. Frequency curves for number of degrees of freedom (DF) = 1, 2, 5

We can easily show

$$\mathbb{E}\{\chi^2\} = \text{mean} = n,$$
$$\text{var}\{\chi^2\} = 2n,$$
$$\alpha_\nu = \nu\text{th moment} = \overline{(\chi^2)^\nu} = n(n+2)\cdots(n+2\nu-2). \tag{6.9}$$

We will show in the next chapter that as $n \to \infty$, $(\chi^2 - n)/\sqrt{2n} \to$ normal distribution with mean $= 0$ and variance $= 1$. For $n \geq 30$, the normal approximation is fairly accurate.

6.3 *F* Distribution

Suppose we are faced with the question of which of two distributions a set of measurements fits best. For this problem, it is convenient to look at the ratio of the chi-square values for each hypothesis. The F distribution enters naturally into this problem.

Let $y_1, \ldots, y_m$ and $w_1, \ldots, w_n$ be mutually independent and normal $(0,1)$. Normal $(0,1)$ is an abbreviation for normal with mean 0 and variance 1. The F distribution is then defined as

$$F = \frac{\sum_{\nu=1}^{m} y_\nu^2 / m}{\sum_{\mu=1}^{n} w_\mu^2 / n}. \tag{6.10}$$

It can be shown that the density function for F is

$$f_{m,n}(F)\, dF = \frac{\Gamma\left(\frac{m+n}{2}\right)\left(\frac{m}{n}F\right)^{m/2-1} dF}{\Gamma\left(\frac{n}{2}\right)\Gamma\left(\frac{m}{2}\right)\left(\frac{mF}{n}+1\right)^{(m+n)/2}}. \tag{6.11}$$

$$\mathbb{E}\{F\} = \frac{n}{n-2} \quad \text{for } n > 2, \tag{6.12}$$

$$\text{var}\{F\} = \frac{2n^2(n+m-2)}{(m(n-2)^2(n-4))} \quad \text{for } n > 4. \tag{6.13}$$

Alternatively, if we let

$$x = \frac{\sum_\nu y_\nu^2}{\sum_\mu w_\mu^2}, \tag{6.14}$$

then

$$f(x)\, dx = \frac{\Gamma\left(\frac{m+n}{2}\right) x^{m/2-1}\, dx}{\Gamma\left(\frac{m}{2}\right)\Gamma\left(\frac{n}{2}\right)(x+1)^{(m+n)/2}}. \tag{6.15}$$

When we compare two different distributions to a measurement, a very large or very small value of F can indicate which distribution is a better fit to the data. However, we should also check the individual chi-squared distributions to see whether either distribution is a reasonable fit to the data.

6.4   Student's Distribution

We often take samples from a distribution we believe to be approximately normal, but whose variance is unknown. Student's distribution is often useful for this problem as it effectively estimates the variance from the data.

Let $x, x_1, x_2, \ldots, x_n$ be mutually independent and normal $(0, \sigma^2)$.

$$t = \frac{x}{\sqrt{\frac{1}{n} \sum_{i=1}^{n} x_i^2}}. \tag{6.16}$$

Note that for a normal distribution with 0 mean, $\frac{1}{n} \sum_{i=1}^{n} x_i^2$ is an estimate from the data of σ^2. In Chap. 12, we will note that n can be described as the effective "number of degrees of freedom".

t^2 has the F distribution with $m = 1$. The frequency function for t is

$$f_n(t) = \frac{1}{\sqrt{n\pi}} \frac{\Gamma\left(\frac{n+1}{2}\right)}{\Gamma\left(\frac{n}{2}\right)} \left(1 + \frac{t^2}{n}\right)^{-(n+1)/2}. \tag{6.17}$$

This is known as Student's distribution or the t distribution (Student was a pseudonym for W.S. Gosset, a pioneer statistician who worked for the Guinness Brewery in Dublin.)

$$\mathbb{E}\{x\} = 0 \text{ for } n > 1, \tag{6.18}$$

$$\text{var}\{x\} = \frac{n}{n-2} \text{ for } n > 2. \tag{6.19}$$

For large n Student's distribution approaches the normal distribution.

6.5   The Uniform Distribution

This is the distribution encountered in Worked Problem 2.1. The density function for the uniform distribution is

$$f(x) = \frac{1}{b-a}, \text{ if } a \leq x \leq b, \text{ and } 0 \text{ otherwise.} \tag{6.20}$$

This is a "square wave" distribution, with x uniformly distributed between a and b. The expectation value and variance are given by

$$\mathbb{E}\{x\} = \int_a^b \frac{x}{b-a} dx = \frac{1}{2}(a+b), \tag{6.21}$$

$$\text{var}\{x\} = \int_a^b [x - \frac{1}{2}(a+b)]^2 \frac{1}{b-a} dx = \frac{1}{12}(b-a)^2. \tag{6.22}$$

6.6 The Log-Normal Distribution

In this distribution, $\log x$ is distributed according to a normal distribution.

$$f(x) = \frac{1}{\sqrt{2\pi\sigma^2}} \frac{1}{x} e^{-(\log x - \mu)^2 / 2\sigma^2}. \tag{6.23}$$

The $1/x$ is present since $d(\log x) = dx/x$. It can be shown that the expectation value and variance of this distribution are not simply μ and σ^2, but are given as

$$\mathbb{E}\{x\} = e^{[\mu + (1/2)\sigma^2]}, \tag{6.24}$$

$$\text{var}\{x\} = e^{(2\mu + \sigma^2)}(e^{\sigma^2} - 1). \tag{6.25}$$

If one is considering the resolution of an instrument, which has a number of sources of error, each contributing a given percentage of the resolution, then the resolution may have a log-normal distribution.

As another example of this distribution consider photomultiplier signals. A photomultiplier is a device for converting weak photon signals into electrical signals. Incoming photons are converted into electrons using the photoelectric effect. The electrons are accelerated and smash into a metal plate, each electron knocking out several secondary electrons. These secondary electrons are accelerated and strike another plate. The process is repeated many times; common photomultipliers often having 14 stages.

Suppose the amplification at stage i is a_i. Let $n_0 = a_0$ be the initial number of electrons produced from the photons. The final number of electrons after k stages is $n_k = \prod_{i=0}^{k} a_i$; $\log n_k = \sum_{i=0}^{k} \log a_i$. We will show in Chap. 10 that the central limit theorem indicates that the sum of k variables often approaches a normal distribution with variance equal to the sum of the individual variances for large k. The conditions of the theorem can be shown to apply here. Then n_k is approximately distributed according to a log-normal distribution.

Suppose, at stage i, an individual electron knocks out a_i electrons with a variance of σ_i^2. If there are n_{i-1} incoming electrons at stage i, then there are n_{i-1} independent trials. The multiplication factor for the n_{i-1} electrons together then has a variance of σ_i^2/n_{i-1}. As an approximation use the average n_{i-1} to determine the variance. Because of the denominator of n_{i-1}, for most stages the variance of a_i is small compared to a_i. Since $\Delta \log a_i = \Delta a_i/a_i$, the variance of $\log a_i$ is approximately equal to the variance of a_i divided by a_i^2.

Let the variance in the number of electrons after the initial photoelectric effect be σ_0/n_{-1}, where n_{-1} is the number of initial photons. The variance of $\log n_k$ is $\approx \sum_{i=0}^{k} \sigma_i^2/(n_{i-1}a_i^2)$. Since we are taking the average value for the n_{i-1} in the denominators, the variance of $\log n_k$ becomes $\sum_{i=0}^{k} \sigma_i^2/(a_i n_{-1} \prod_{j=0}^{i} a_j)$.

6.7   The Breit–Wigner (Cauchy) Distribution

The density function of the Cauchy distribution is

$$f(E)\,dE = \frac{1}{\pi} \frac{\Gamma/2\,dE}{(E - E_0)^2 + (\Gamma/2)^2}. \tag{6.26}$$

This distribution is the often used distribution in physics that describes an energy spectrum near a resonance. Γ is the full width of the resonance if measured at one-half the maximum height of the resonance (FWHM). The distribution function is

$$F = \frac{1}{\pi}\frac{\Gamma}{2} \int_{-\infty}^{E} \frac{dE'}{(E' - E_0)^2 + (\Gamma/2)^2}. \tag{6.27}$$

In physics, the lower limit is not $-\infty$, but the minimum allowable energy. This is usually so far from the peak that the normalization discrepancy is small.

This distribution describes a wealth of phenomena. It describes the energy spectrum of an excited state of an atom or molecule, as well as an elementary particle resonant state. It can be shown quantum mechanically that whenever one has a state that decays exponentially with time, the energy width of the state is described by the Cauchy distribution.

The F above can be integrated to give

$$F = \frac{1}{\pi} \tan^{-1}\left(\frac{E - E_0}{\Gamma/2}\right) + \frac{1}{2}. \tag{6.28}$$

What is $\mathbb{E}\{E\}$?

$$\mathbb{E}\{E\} = \int Ef(E)\,dE = \frac{1}{\pi}\frac{\Gamma}{2}\int_{-\infty}^{\infty}\frac{E\,dE}{(E-E_0)^2 + (\Gamma/2)^2}$$

$$= \frac{1}{\pi}\frac{\Gamma}{2}\int_{-\infty}^{\infty}\frac{[(E-E_0)+E_0]}{(E-E_0)^2 + (\Gamma/2)^2}\,dE. \tag{6.29}$$

The $E - E_0$ term $= \dfrac{1}{2}\dfrac{1}{\pi}\dfrac{\Gamma}{2}\,\log\left[(E-E_0)^2 + \left(\dfrac{\Gamma}{2}\right)^2\right]_{-\infty}^{\infty} =$ indeterminate!

$$\tag{6.30}$$

$$\text{The } E_0 \text{ term } = E_0 = \text{finite.} \tag{6.31}$$

Thus, $\mathbb{E}\{E\}$ is indeterminate. (If we take a symmetric limit, i.e., the limit as $L \to \infty$ of $\int_{-L-E_0}^{L+E_0} = E_0$.)

The higher moments also do not exist. This is a symptom of serious problems with this distribution. We will discuss this further in Chap. 11.

6.8 Beta Distribution

The density function for the beta distribution is

$$f(x) = \frac{\Gamma(m+n)}{\Gamma(m)\Gamma(n)}x^m(1-x)^n dx. \tag{6.32}$$

For this distribution x is bounded $0 \leq x \leq 1$. The expected value of x and its variance are given by

$$\mathbb{E}(x) = m/(m+n); \quad var(x) = \frac{mn}{(m+n)^2(m+n+1)}. \tag{6.33}$$

The uniform distribution is a beta distribution with $m = n = 1$.

The Γ functions can get quite large, even though $F(m, n)$ is of moderate size. Stirling's approximation for the factorial function can be useful here. We use the full accuracy (second order) Stirling's approximation here. It is best to use the logarithms of Γ functions and then exponentiate afterward. The approximation is quite good. Even for $m = 1$, $n = 2$ the result is accurate to three decimal places. A code fragment is

```
#include <cmath> // allows access to functions like log and exp
#include <vector> // needed to use vectors
#include <random> // random number generation
#include <iostream>
```

```
#include <fstream>
#include <math.h>
int main() {
}
// For beta distribution
double lnstirl(double m) {
double twopi = 6.2831853
return m * (log(m) - 1) + .5 * log(twopi * m) + 1/(12 * m);
}
double stirlmn(double m, double n) {
double mp;
double np;
// For m=1 and n =1 and noting Gamma (0) = Gamma (1)=1, const = 1.
// stirlmn should be set to 1.  Not valid approx if m,n<1, and
// probably not accurate if m=1 and n not to far from 1 (it is good
// to 3 decimal places if m=1 and n=2.)
if (m == 1 || n == 1) {
return max(n, m);
}
else {
mp = m - 1;
np = n - 1;
return exp(lnstirl(mp + np + 1) - lnstirl(mp) - lnstirl(np));
}
}
```

Note that this fragment treats m and n as real not as integer. One advantage of using Stirling's approximation is that an expanded beta distribution is available which interpolates between integral values in m and n. This can be useful when trying to match a beta distribution with experimental data. The dashed line in Fig. 6.6 has non-integral m, n.

6.9   Exponential Distributions

6.9.1   Exponential Distribution

We have noted that this distribution is a particular limit of the gamma distribution. This distribution is a very important distribution when studying radioactive atoms. Suppose there is a sample of decaying radioactive atoms and that the probability of decay of an individual atom is constant per unit time, λdt. If there are N atoms left, the probability of some atom decaying over the next dt is $N\lambda dt$.

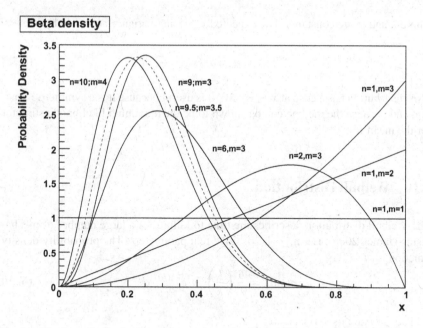

Fig. 6.6 The beta distribution for various values of m and n. Note that the dashed line is an interpolation to non-integral m, n values

$$\frac{dN}{dt} = -N\lambda dt; \quad \frac{dN}{N} = -\lambda dt. \tag{6.34}$$

The solution to this equation with the boundary condition that there are N_0 atoms at $t = 0$ and the probability density function is

$$N = N_0 e^{-\lambda t}; \quad f(t) = \lambda e^{-\lambda t}. \tag{6.35}$$

This is the exponential distribution. The expectation value and variance are given by

$$\mathbb{E}(t) = \frac{1}{\lambda}; \quad var(t) = \frac{1}{\lambda^2}. \tag{6.36}$$

6.9.2 Double Exponential Distribution

The probability density function for the double exponential distribution, also known as the Laplace distribution is

$$p(x)dx = \frac{\lambda}{2}e^{-\lambda|x-\mu|}; \tag{6.37}$$

where λ and μ are constants. The expected value and variance are given by

$$\mathbb{E}(x) = \mu; \; var(x) = \frac{2}{\lambda^2}. \tag{6.38}$$

This distribution has a cusp at $x - \mu = 0$. It is useful because it is a symmetric function and because the median can be shown to be the maximum likelihood estimator for the mean μ.

6.10   Weibull Distribution

The Weibull distribution describes the time to failure for a large number of mechanisms (James 2006). Let η, σ be positive real parameters. The probability density function for $t \geq 0$ is

$$p(t)dt = \frac{\eta}{\lambda} \left(\frac{t}{\lambda} \right)^{\eta-1} e^{-[(\frac{t}{\lambda})^{\eta}]} dt. \tag{6.39}$$

The $\mathbb{E}(t)$ and its variance are:

$$\mathbb{E}(t) = \lambda \Gamma \left(\frac{1}{\eta} + 1 \right); \; var(t) = \lambda^2 \left[\Gamma \left(\frac{2}{\eta} + 1 \right) - \left[\Gamma \left(\frac{1}{\eta} + 1 \right) \right]^2 \right], \tag{6.40}$$

where Γ is the gamma function. It interpolates between the exponential distribution ($\eta = 1$) and the Rayleigh distribution ($\eta = 2$) with $\lambda = \sqrt{2}\sigma$. The Rayleigh distribution is the subject of Exercise 6.1, where you will show it to be the probability of a first failure of a device at time t if it has performed without failure up to that time.

In this chapter, we have introduced the normal and other continuous distributions that are often encountered in applications. They often appear as the limit of other distributions and the normal distribution is the most important single distribution in this subject. The chi-square distribution will be central when we consider statistics questions, questions of estimating parameters from data. The Cauchy distribution, which appears in physics as the Breit–Wigner distribution, is a peculiar distribution for which the higher moments do not exist. This will lead to strange limiting behavior, which we will discuss in Chap. 10.

6.11   Worked Problems

WP6.1 Show that for the normal distribution

$$f = \frac{1}{\sqrt{2\pi\sigma^2}} e^{-x^2/2\sigma^2}, \tag{6.41}$$

the parameter σ^2 is indeed the variance (i.e., show $\overline{x^2} = \sigma^2$). Hint: you might
wish to use the Γ function:

$$\int_0^\infty y^{z-1} e^{-y} \, dy = \Gamma(z), \qquad \Gamma(z) = (z-1)\Gamma(z-1), \qquad (6.42)$$

$$\Gamma\left(\frac{1}{2}\right) = \sqrt{\pi}, \qquad \Gamma(n) = (n-1)!. \qquad (6.43)$$

Answer

$$\overline{x^2} = \frac{1}{\sqrt{2\pi\sigma^2}} \int_{-\infty}^\infty x^2 e^{-x^2/2\sigma^2} \, dx = \frac{1}{\sqrt{2\pi\sigma^2}} \frac{1}{2} \int_{-\infty}^\infty x e^{-x^2/2\sigma^2} 2x \, dx.$$

Note that $2x \, dx = dx^2$. Let $y = x^2/2\sigma^2$.
Note that $\int_{-\infty}^0 dx$ and $\int_0^\infty dx$ both go to $\int_0^\infty dy$.

$$\overline{x^2} = \frac{1}{\sqrt{2\pi\sigma^2}} \frac{2}{2} \int_0^\infty \sqrt{2\sigma^2 y} e^{-y} 2\sigma^2 \, dy$$

$$= \frac{2\sigma^2}{\sqrt{\pi}} \int_0^\infty \sqrt{y} e^{-y} \, dy = \frac{2\sigma^2}{\sqrt{\pi}} \frac{\sqrt{\pi}}{2} = \sigma^2,$$

where we noted

$$\int_0^\infty y^{\frac{1}{2}} e^{-y} \, dy = \Gamma\left(\frac{3}{2}\right) = \frac{1}{2}\Gamma\left(\frac{1}{2}\right) = \frac{\sqrt{\pi}}{2}.$$

Another proof of this problem comes from considering

$$\frac{1}{2\pi\sigma^2} \int (x^2 + y^2) e^{-(x^2+y^2)/2\sigma^2} \, dx \, dy = 2 * \text{variance}.$$

$$\frac{1}{2\pi\sigma^2} \int r^2 e^{-r^2/2\sigma^2} r \, dr \, d\theta = \frac{(2\sigma^2)^2}{2\sigma^2} \int x e^{-x} \, dx,$$

where we noted that the integral over θ gave 2π. But

$$\int_0^\infty x e^{-x} \, dx = x e^{-x}\big|_0^\infty + \int_0^\infty e^{-x} \, dx = 1. \qquad (6.44)$$

Thus, $2 * \text{variance} = 2\sigma^2$.

6.12　Exercises

6.1 Suppose the probability that an electronic component fails to function in the time interval $(t, t + dt)$ is $\phi(t)\, dt$ provided the component has functioned properly up to time t. Given $\int_0^\infty \phi(\tau)\, d\tau = \infty$, show that

$$p(t) = \phi(t)e^{-\int_0^t \phi(\tau)\, d\tau} \tag{6.45}$$

is the probability density function (for positive t) for the time at which the component first fails to function. [If $\phi(t) = t/\alpha^2$, then $p(t) = t/\alpha^2 e^{-t^2/2\alpha^2}$ is known as the Rayleigh probability density function.]

6.2 You are measuring a quantity and are uncertain of the error. You make three measurements obtaining 1.0, 2.0, 1.0. Estimate the variance. The next day you come back to your apparatus and measure 7.0. You suspect that something has changed, but worry that it might be only that the variance is larger than you thought. Using all four measurements, use the t distribution and look at the value of t for the last measurement. Compare the value of the density function for $t = 0$ to the density function value for your t. Are you reasonably certain that something has changed in your apparatus?

6.3 Suppose we make five measurements of a quantity we know to be normally distributed with mean 0.05. We obtain 0.041, 0.064, 0.055, 0.046, 0.060. Estimate the variance of the distribution. Suppose we know that the variance is, in fact, 1.0×10^{-4}. What is the value of χ^2 for this set of measurements and, using Fig. 6.5, to what value of the χ^2 density function does this correspond?

6.4 We have seen that the binomial distribution approaches the normal distribution for large n. Compare the exact calculation of the probability of having 20 heads out of 40 tosses of a fair coin with the approximation using the normal distribution.

6.5 Suppose we have N books numbered 1 to N on a shelf and randomly select two of them located at j_1 and j_2. The probability for the sum $j_1 + j_2$ is called the triangle distribution. The distribution is given by:

$$p(x) = \frac{1 - |x - \mu|}{\Gamma} \quad \text{for } \mu - \Gamma < x < \mu + \Gamma \text{ and } 0 \text{ otherwise.} \tag{6.46}$$

The mean and variance are given by:

$$\mathbb{E}(x) = \mu; \quad var(x) = \frac{\Gamma^2}{6} \tag{6.47}$$

Derive the expression for the mean value.

Reference

James F (2006) Statistical methods in experimental physics, 2nd edn. World Scientific Publishing Co. Pte. Ltd, Singapore

Chapter 7
Generating Functions and Characteristic Functions

Abstract Convolutions, compound probability, generating functions, and characteristic functions are of considerable use in theoretical probability, i.e., proving probability theorems. They are also of use to us when we wish to put distributions together. Consider $x = x_1 + x_2 + \cdots + x_n$, where x_1 is distributed according to one distribution, x_2 according to another, etc. Sometimes the number of distributions is not fixed, but distributed according to some random distribution also. In the present chapter, we will consider several examples of applications of this kind.

7.1 Convolutions and Compound Probability

Let $x = x_1 + x_2$, where x_1 and x_2 are independent random variables distributed according to distributions F_1 and F_2. The distribution (F) of x is called the convolution of the distributions of x_1 and x_2. For discrete variables with integral possible values, we let

$$Pr[x_1 = j] = a_j, \tag{7.1}$$

$$Pr[x_2 = k] = b_k. \tag{7.2}$$

Then, we have

$$Pr[x_1 + x_2 = t] \equiv c_t = a_0 b_t + a_1 b_{t-1} + \cdots + a_t b_0. \tag{7.3}$$

c is called the convolution of a and b or the composition of a and b and is denoted by $c = a \star b$. $c_t = (a \star b)_t = Pr[x_1 + x_2 = t]$.

For continuous variables, we have

$$F[t = x_1 + x_2] = \int_{-\infty}^{+\infty} F_1(t - x) \, dF_2(x) = \int_{-\infty}^{+\infty} F_2(t - x) \, dF_1(x). \tag{7.4}$$

(Using generalized integrals, this would include the discrete case as well.) We say $F = F_1 \star F_2$. F is the convolution or composition of F_1 and F_2.

© Springer Nature Switzerland AG 2020

B. P. Roe, *Probability and Statistics in the Physical Sciences*,
Undergraduate Texts in Physics,
https://doi.org/10.1007/978-3-030-53694-7_7

For completely continuous distributions, if at least one of the density functions is bounded for all x, then we can derive a relation for the density function $f(t)$ of the convolution in terms of the density functions $f_1(x_1)$ and $f_2(x_2)$ of the components:

$$f(t) = \int_{-\infty}^{+\infty} f_1(t-x)f_2(x)\,dx. \tag{7.5}$$

For either discrete or continuous probabilities, we can easily show

$$F_1 \star (F_2 \star F_3) = (F_1 \star F_2) \star F_3 = (F_2 \star F_3) \star F_1. \tag{7.6}$$

Hence, we can simply speak of this as $F_1 \star F_2 \star F_3$. This is intuitively evident when we realize this just corresponds to the sum of three independent random variables. The extension to any number is immediate.

Suppose we have a sum of N independent random variables from the *same* distribution.

$$s_N = x_1 + x_2 + \cdots + x_N. \tag{7.7}$$

Clearly, $F(s_N) = F \star F \star F \star \cdots \star F$ (N times). This is denoted $\{F\}^{N^*}$ and, for the discrete case, $\{a\}^{N^*}$. Consider now the case where N is not fixed. Suppose we have

$$Pr[N = n] = g_n. \tag{7.8}$$

This is called a compound probability. If the x_i are discrete, integer-valued variables, then

$$Pr[s_N = j] = \sum_{N=0}^{\infty} g_N \{a\}^{N^*}. \tag{7.9}$$

Exercise 5.3 was an example of a compound probability. There we imagined that a high-energy interaction led to a Poisson distribution in the number of produced pion pairs and that each pair had a probability p of being a $\pi^+\pi^-$ and $1-p$ of being $2\pi^0$. We asked what the probability was of having n charged pairs. We will shortly see a simpler method of solving this problem than was available before.

7.2   Generating Functions

Generating functions are defined for discrete distributions of random variables with integral, non-negative possible values. Let

$$Pr[x = j] = p_j, \tag{7.10}$$

$$Pr[x > j] = r_j, \qquad r_j = \sum_{k=j+1}^{\infty} p_k = 1 - \sum_{k=0}^{j} p_k. \qquad (7.11)$$

Generating functions are functions of a variable s whose power series expansion has the desired probabilities as coefficients. Define the generating functions:

$$P(s) = \sum_{j=0}^{\infty} p_j s^j, \qquad (7.12)$$

$$R(s) = \sum_{j=0}^{\infty} r_j s^j. \qquad (7.13)$$

Using the relation between p_j and r_j, the following theorems can be proven relatively easily:

Theorem 7.1

$$\text{If } -1 < s < 1, \qquad R(s) = \frac{1 - P(s)}{1 - s}. \qquad (7.14)$$

Theorem 7.2

$$p_k = \frac{1}{k!} \frac{d^k P}{ds^k}\Big|_{s=0}, \qquad (7.15)$$

and, therefore, the generating function determines the distribution. Note that $P(1) = 1$ as it is just the sum of the probabilities which add to 1.

Theorem 7.3

$$P'(1) \equiv \frac{dP}{ds} = \sum_{k=0}^{\infty} k p_k = < k > \equiv < x >, \text{ Thus :} \qquad (7.16)$$

$$\mathbb{E}\{x\} \equiv \sum_{k=0}^{\infty} k p_k = \sum_{k=0}^{\infty} r_k = P'(1) = R(1). \qquad (7.17)$$

Equation 7.16 generalizes:

$$< k^m > = \left|\left(s\frac{d}{ds}\right)^m P(s)\right|_{s=1}. \qquad (7.18)$$

$$If \; \mathbb{E}\{x^2\} \equiv \sum_{j=1}^{\infty} j^2 p_j \; is \; finite, \; then, \; if \; we \; do \; not \; use \; Eq. \; 7.18 \tag{7.19}$$

$$\mathbb{E}\{x^2\} = P''(1) + P'(1) = 2R'(1) + R(1),$$

$$var(x) = P''(1) + P'(1) - [P'(1)]^2 = 2R'(1) + R(1) - [R(1)]^2.$$

The variance will be infinite if and only if $P''(s) \to \infty$ as $s \to 1$.

From extensions of Theorem 7.3 we see that the moments of a distribution are given in terms of derivatives of P (or R) evaluated at $s = 1$. The generating functions, therefore, can be used to find all finite moments of a distribution.

Theorem 7.4 For convolutions, if $p = a \star b$ and if the generating functions are P, P_a, P_b, respectively, then

$$P(s) = P_a(s)P_b(s). \tag{7.20}$$

Theorem 7.5 Suppose we are putting together a variable number N of distributions the ith of which has probabilities $p_k^{(i)}$. Since the individual probabilities $p_k^{(i)}$ are independent, the probability that they take a particular set of values k_i is simply the product $\prod_i p_{k_i}$ of their individual probabilities. The probability $\pi(j)$ that their sum adds up to j is

$$\pi(j) = \sum_{k_1=0}^{\infty} \cdots \sum_{k_m=0}^{\infty} \delta\left(j, \sum_i k_i\right) \prod_{i=1}^{m} p_{k_i}^{(i)} \tag{7.21}$$

Theorem 7.6 The generating function h(x) for $\pi(j)$ can be found by summing over j.

$$h(x) = \sum_{j=0}^{\infty} \pi(j)x^j = \sum_{j=0}^{\infty} x^j \sum_{k_1=0}^{\infty} \cdots \sum_{k_m=0}^{\infty} \delta\left(j, \sum_i k_i\right) \prod_{i=1}^{m} p_{k_i}^{(i)} = \prod_{i=1}^{m} P^i(x); \tag{7.22}$$

where $P^i(x) = \sum_{k=0}^{\infty} p_k^{(i)} x^k$. If the samples are all drawn from the same distribution,

$$h(x) = [P(x)]^m \tag{7.23}$$

Theorem 7.7 In Theorem 7.6, we found the generating function h(x) for m independent distributions. If we have a variable number m of distributions and if P_g is the generating function for the number of distributions m, we can then show that

$$P(x) = P_g[h(x)]. \tag{7.24}$$

Table 7.1 Generating functions for various distribution

Distribution	Generating function
Binomial	$P(s) = (q + ps)^n$
Poisson	$P(s) = e^{-\lambda + \lambda s}$
Geometric	$P(s) = \dfrac{p}{1 - qs}$
Multinomial	$P(s_2, s_3, \ldots, s_k) = (p_1 + p_2 s_2 + \cdots + p_k s_k)^N$
Negative binomial	$P(s) = \left(\dfrac{ps}{1 - (1-p)s}\right)^m$

Theorem 7.8 *We consider next a result from network theory. Build a network such that the probability of a node having k spokes to other nodes is p_k. The mean value of k is $< k >= \sum_{i=0}^{\infty} k p_k$ and the generating function is $P(x) = \sum_{k=0}^{\infty} p_k x^k$. Pick a node with k spokes and ask what is the average number of spokes for the set of k nodes connected to the spokes of the selected node. The remarkable result is that that average is $q_k = (k + 1)p_{k+1}/ < k >$ which is larger than k. "Your friends have more friends than you have!" q_k is called the excess degree distribution. The generating function for q_k is $Q(x)$. We will leave it for the exercises to prove that*

$$Q(x) \equiv \sum_{k=0}^{\infty} q_k x^k = \frac{P'(x)}{P'(1)} \tag{7.25}$$

Some generating functions for common distributions are given in Table 7.1. In the table, $q = 1 - p$. The geometric distribution is the number of trials up to the first success in a string of Bernoulli trials, $Pr[x = j] = q^j p$.

It is easily seen from the above that if $x = x_1 + x_2$ and if x_1 and x_2 are both binomial with the same p and $n = n_1$, n_2, respectively, then x is binomially distributed with that p and $n = n_1 + n_2$. The sum of two binomial distributions is a binomial distribution.

Similarly if x_1 and x_2 are Poisson distributed with $\lambda = \lambda_1$, λ_2, respectively, $x = x_1 + x_2$ is Poisson distributed with $\lambda = \lambda_1 + \lambda_2$. The sum of two Poisson distributions is a Poisson distribution.

As an example of compound probability, let us consider the pion pair problem given as Exercise 5.3. Suppose N is Poisson distributed with parameter λ, and x_i has probability p of being 1 and $q = 1 - p$ of being 0. Then each term x_i can be regarded as binomial with $n = 1$. Then $P_{s_N}(s) = P_g(P_a(s)) = e^{-\lambda + \lambda(q + ps)} = e^{-\lambda p + \lambda ps}$. This is the generating function for a Poisson distribution with parameter λp and, thus, we obtain the probability of n charged pion pairs.

As a second example, imagine we are tossing a die until we get a 6. En route, how many 1's do we get? In general, let g be a geometrical distribution. For the geometrical distribution, we let $\gamma = 1 - p$ to avoid confusion. x_1 has probability p ($= 1/5$) of being 1 and q of being 0. We have

$$P_{s_N}(s) = \frac{1 - \gamma}{1 - \gamma(q + ps)} = \frac{1 - \alpha}{1 - \alpha s}, \tag{7.26}$$

where $\alpha = (\gamma p)/(1 - \gamma q)$. Thus, the result is a new geometrical distribution.

For a final example, we consider Poisson trials. These are a generalization of Bernoulli trials. They are a series of N mutually independent trials, the kth of which has probability p_k for $x = 1$ and $q_k = 1 - p_k$ for $x = 0$. This differs from Bernoulli trials in that we allow the probability to vary from trial to trial. We let $s_N = x_1 + x_2 + \cdots + x_N$. Then the generating function $P_{s_N} = (q_1 + p_1 s)(q_2 + p_2 s) \cdots (q_N + p_N s)$. Suppose as $N \to \infty, p_1 + p_2 + \cdots + p_N \to \lambda$, where λ is a finite constant, and $p_k \to 0$ for each k. Consider $\log P_{s_N}(s) = \Sigma_{k=1}^N \log(q_k + p_k s) = \Sigma_{k=1}^N \log\{1 - p_k(1 - s)\}$. As $p_k \to 0, \log\{1 - p_k(1 - s)\} \to -p_k(1 - s)$. Thus, $\log P_{s_N}(s) \to -\lambda(1 - s)$ and, hence, the distribution of s_N approaches a Poisson distribution with parameter λ. This is an example of a central limit theorem, which we will treat further in Chap. 11.

7.3 Characteristic Functions

Characteristic functions (moment generating functions) are a generalization of generating functions to continuous distributions. When a distribution is continuous and the density function exists, characteristic functions are simply the Fourier transforms of the density functions. We define the characteristic function (c.f.) as

$$\phi(s) = \mathbb{E}\{e^{isx}\} = \int_{-\infty}^{\infty} e^{isx}\, dF(x). \qquad (7.27)$$

For discrete variables, this becomes $p(0) + e^{is1}p(1) + e^{is2}p(2) \cdots$. If we define $z \equiv e^{is}$, then the characteristic function becomes $p(0) + zp(1) + z^2 p(2) \cdots$. This is essentially the generating function (except that z is complex and the s for generating functions is real). We can easily see that the c.f. of $g(x)$ is $\mathbb{E}\{e^{isg(x)}\}$ and

$$\mathbb{E}\{e^{is(ax+b)}\} = e^{bis}\phi(as), \qquad (7.28)$$

$$\mathbb{E}\{e^{is(x-m)/\sigma}\} = e^{-mis/\sigma}\phi\left(\frac{s}{\sigma}\right). \qquad (7.29)$$

We quote some theorems below concerning characteristic functions.

Theorem 7.9 *The characteristic function always exists.*

Theorem 7.10

$$\left.\frac{d^n\phi(s)}{ds^n}\right|_{s=0} = i^n\overline{x^n}. \qquad (7.30)$$

Also, the cumulant generating function is defined as

$$K(s) = \log \phi(s) = \sum_{j=1}^{\infty} \frac{\chi_j}{j!}(is)^j, \qquad (7.31)$$

where the quantities χ_j are called the semi-invariants or cumulants of the distribution. We have

$$\chi_1 = m,$$
$$\chi_2 = \sigma^2,$$
$$\chi_3 = \mu_3, \tag{7.32}$$
$$\chi_4 = \mu_4 - 3\sigma^4,$$
$$\chi_5 = \mu_5 - 10\sigma^2\mu_3.$$

For the normal distribution, all χ_i $(i \geq 3)$ are 0.

Theorem 7.11 *The characteristic function uniquely determines the distribution in all continuity intervals.*

Theorem 7.12 *The characteristic function, ϕ, of the convolution of two independent random variables $(x = x_1 + x_2)$ is the product of the characteristic functions of each variable $\phi(s) = \phi_1(s)\phi_2(s)$. This is just the well-known convolution theorem for Fourier transforms.*

Theorem 7.13

$\chi_m(x_1 + x_2) = \chi_m(x_1) + \chi_m(x_2)$. *This is clear from Theorems 7.2 and 7.4.* (7.33)

Theorem 7.14 *If as $n \to \infty$, the characteristic function of $x_n \to$ the characteristic function of y, then the distribution function for $x_n \to$ the y distribution function.*

Some characteristic functions for common distributions are given in Table 7.2.

Let us consider the convolution of two χ^2 distributions in n_1 and n_2 degrees of freedom. It is clear from the definition of the distribution, but also from Theorems

Table 7.2 Characteristic functions for various distributions

Distribution	Characteristic function
Binomial	$\phi(s) = (pe^{is} + q)^n$
Poisson	$\phi(s) = e^{\lambda(e^{is}-1)}$
Normal	$\phi(s) = e^{ims - s^2\sigma^2/2}$
χ^2 (n d.f.)	$\phi(s) = \mathbb{E}\{e^{is\chi^2}\} = (1 - 2is)^{-n/2}$
Exponential $f(x) = [(1/\lambda)e^{-x/\lambda}]$	$1/(1 - is/\lambda)^{-1}$
Breit–Wigner	$\phi(s) = e^{-iE_0 s - (\Gamma/2)|s|}$
Gamma $f(x) = [\mu e^{-\mu x}(\mu x)^{\alpha-1}/(\alpha-1)!]$	$(1 - is/\mu)^{-\alpha}$
Negative binomial	$\phi(s) = [(e^{-is} - q)/p]^{-m}$
Uniform (a to b)	$\phi(s) = \frac{e^{isb} - e^{isa}}{is(b-a)}$
Triangle	$\phi(s) = \frac{2e^{isu}}{s^2\Gamma^2}[1 - \cos(s\Gamma)]$
Double exponential	$\phi(s) = isu + \frac{\lambda^2}{\lambda^2 + s^2}$

7.3 and 7.4 and the χ^2 characteristic function, that the convolution is χ^2 in $n_1 + n_2$ degrees of freedom.

Next suppose $n \to \infty$ in a χ^2 distribution with n degrees of freedom. Consider the variable $t = (\chi^2 - n)/\sqrt{2n}$. (Remember, for the χ^2 distribution the mean is n and the variance $2n$.) The characteristic function for t is

$$\phi(s) = e^{-is\sqrt{n/2}}(1 - is\sqrt{2/n})^{-n/2} = (e^{is\sqrt{2/n}}(1 - is\sqrt{2/n}))^{-n/2}. \tag{7.34}$$

For fixed s, we choose n large and expand the exponential obtaining $(1 + s^2/n + \theta(2/n)^{3/2}|s|^3)^{-n/2}$, where $\theta \le 1$. As $n \to \infty$, this approaches $e^{-s^2/2}$, since this form is essentially the form of the basic definition of e. Hence, the χ^2 distribution is asymptotically normal. (In practice, if $n \ge 30$, then the distribution is essentially normal.) This is another example of a distribution approaching the normal distribution. In Chap. 11, we shall consider the general criteria under which distributions will approach the normal distribution (central limit theorems).

The characteristic function can be used to find the mean and variance of distributions. Consider, for example, the log-normal distribution. Set $y = \log x$. Then the characteristic function can be considered as $\mathbb{E}\{e^{ise^y}\}$, using the normal distribution density function for $f(y)$. If one expands the exponential in powers of s, the first few terms are easily integrable and can be compared with the expressions for the semi-invariants to yield the mean and variance. For distributions whose characteristic functions are known, simply taking the derivatives of the characteristic functions and using Eq. 7.30 enables us to find the mean and variance.

In this chapter, we have introduced characteristic functions and generating functions. These concepts are of use in probability theory and also when dealing with convolutions of random variables or with compound probability. We found that either the characteristic or the generating function of a distribution completely determines the distribution and that the various moments of the distribution are just combinations of derivatives of these functions. The generating function (characteristic function) of the convolution of two random variables is just the product of the two individual generating (characteristic) functions.

7.4 Exercises

7.1 If x_i is a set of independent measurements, let $x_{AV} = \Sigma x_i/n$. Prove that for the Breit–Wigner distribution the distribution of x_{AV} is the same as that of x_i. (Recall that for a normal distribution, the distribution of x_{AV} has standard deviation decreasing with increasing n.) What is the σ of x_{AV}? Hint: Use the characteristic function.

7.2 Suppose, for X, a discrete random variable, $f_n = Pr(X = n)$ has generating function $F(s)$. Let $g_n = Pr(X > n)$ and have generating function $G(s)$. Show that $G(s) = (1 - F(s))/(1 - s)$.

7.3 Prove the generating functions Eqs. 7.14, 7.18.

7.4 Prove Eq. 7.20.

7.5 Prove Eq. 7.22.

$$h(x) = \sum_{j=0}^{\infty} \pi(j)x^j = \sum_{j=0}^{\infty} x^j \sum_{k_1=0}^{\infty} \cdots \sum_{k_m=0}^{\infty} \delta(j, \sum_i k_i) \prod_{i=1}^{m} p_{k_i}^{(i)} = \prod_{i=1}^{m} P^i(x);$$

where $P^i(x) = \sum_{k=0}^{\infty} p_k^{(i)} x^k$. The probability $\pi(j)$ was

$$\pi(j) = \sum_{k_1=0}^{\infty} \cdots \sum_{k_m=0}^{\infty} \delta\left(j, \sum_i k_i\right) \prod_{i=1}^{m} p_{k_i}^{(i)}$$

7.6 Prove Theorem 7.8 for generating functions (Eq. 7.25).

$$Q(x) \equiv \sum_{k=0}^{\infty} q_k x^k = \frac{P'(x)}{P'(0)},$$

where $q_k = (k+1)p_{k+1}/ < k >$.

7.7 Find the mean and standard deviation of the χ^2 distribution using the characteristic function.

Chapter 8
The Monte Carlo Method: Computer Simulation of Experiments

Abstract Many times we wish to simulate the results of an experiment by using a computer and random variables, using the pseudorandom number generators available on computers. This is known as Monte Carlo simulation. This is often done because the experiment is very complicated and it is not practical to analytically summarize all of the different effects influencing our results. We first try to generate a set of representative simulated events. In this chapter, we will examine techniques for generating random events according to their expected probability. Afterward, these events are often run through a representation of the experimental detection apparatus to give the final expected distributions. Since this latter process is very apparatus specific, we will not treat this part of the problem. The Monte Carlo method is necessarily connected with computers, and we will introduce some simple computer problems here

8.1 Using the Distribution Inverse

Most computer libraries have a pseudorandom number generator capable of generating pseudorandom numbers, R, uniformly distributed between 0 and 1. We wish to use these to generate an arbitrary distribution.

We use the integral probability, i.e., the distribution function, F. $P = F(x)$ and P varies from 0 to 1 uniformly. That is, 10% of the events fall between $P = 0$ and $P = 0.1$ and another 10% between $P = 0.6$ and $P = 0.7$. This follows from the definition of F. Hence, we choose $R = F(x)$. Then $x = F^{-1}(R)$ gives us x. This set of x is distributed with distribution function F.

For example, suppose we wish to find a set of numbers distributed according to an exponential distribution with density function:

$$f = \frac{e^{-t/\tau}}{\tau}.$$

This might be the distribution of decay times for a single radioactive particle. The distribution function is determined from

© Springer Nature Switzerland AG 2020

B. P. Roe, *Probability and Statistics in the Physical Sciences*,
Undergraduate Texts in Physics,
https://doi.org/10.1007/978-3-030-53694-7_8

$$F = \int_0^t f \, dt = 1 - e^{-t/\tau}.$$

Thus,

$$1 - F = e^{-t/\tau}.$$

If F is uniformly distributed between 0 and 1, $1 - F$ is also. Hence, we choose $R = e^{-t/\tau}$ or $t = -\tau \log R$. This gives us a set of pseudorandom numbers t with exponential distribution.

For the Breit–Wigner distribution, we found in Chap. 6 that

$$F = \frac{1}{\pi} \tan^{-1}\left(\frac{E - E_0}{\Gamma/2}\right) + \tfrac{1}{2}. \tag{8.1}$$

If we let $F = R$ and solve for E, we get

$$E = E_0 + (\Gamma/2) \tan \pi (R - \tfrac{1}{2}), \tag{8.2}$$

which will give an appropriate distribution of energies.

For some distributions, numerical integration to provide a look-up table is useful in applying the distribution function. That is, we divide F into regions, and within each region, we find the mean value of the parameter of interest. We store these values in a table. When R is picked, we find its corresponding entry. Often, we do a linear interpolation between values within the table.

8.2 Method of Composition

There are a number of tricks that can aid in generating numbers according to a given distribution. Some of these are discussed in Yost (1985) and more fully in Rubenstein (1981).

The method of composition is often useful. We start by illustrating the discrete case. Consider

$$f(x) = \tfrac{5}{12}[1 + (x - 1)^4], \qquad 0 \le x \le 2.$$

We divide this into a sum of two (normalized) density functions:

$$f_a = \tfrac{1}{2}, \qquad f_b = \tfrac{5}{2}(x - 1)^4,$$
$$f(x) = \tfrac{5}{6} f_a + \tfrac{1}{6} f_b.$$

This is the weighted sum of the two densities with weights $\tfrac{5}{6}$ and $\tfrac{1}{6}$. Instead of one pseudorandom number, we use two. R_1 decides whether to use f_a or f_b, i.e., if $R_1 < \tfrac{5}{6}$, use f_a, and otherwise f_b. We then use the second pseudorandom number to invert f_a or f_b as needed.

$$\text{For } f_a \rightarrow x = 2R_2,$$
$$f_b \rightarrow x = 1 + \text{sign}(2R_2 - 1) \times |2R_2 - 1|^{0.2}.$$

We only need to take the 0.2 power one-sixth of the time. Note the difference between the method of composition and the convolution of variables considered in the last chapter. There we considered $x = x_1 + x_2$. Here, we have only one x for each measurement, but we have the probability of $x = \text{probability}_1(x) + \text{probability}_2(x)$. We are adding probabilities, not variables.

Next we consider the continuous case. We proceed by analogy to the discrete $\frac{5}{6}$ and $\frac{1}{6}$ case, but the $\frac{5}{6}$ and $\frac{1}{6}$ themselves become continuous functions. Consider as an example that we wish to generate numbers according to the density function

$$f(x) = n \int_1^\infty y^{-n} e^{-xy} \, dy, \qquad n \geq 1 \text{ and } x \geq 0.$$

Write

$$f(x) = \int_0^1 g(x|y) \, dF(y).$$

Here F is the variable weight and $g(x|y)$ is the conditional probability that we obtain x, given that y is fixed. Here we take

$$g(x|y) = ye^{-xy},$$
$$dF(y) = \frac{n \, dy}{y^{n+1}}, \qquad 1 < y < \infty.$$

Note

$$f(x) = n \int_1^\infty y^{-n} e^{-xy} \, dy = -\int_{y^{-n}=1}^0 ye^{-xy} \, d(y^{-n}) = \int_0^1 ye^{-xy} \, dF.$$

The strategy will be the same as last time. We will pick a pseudorandom number, R to find y, i.e., $R_1 = F(y)$ or $y = F^{-1}(R_1)$. We will then pick R_2 and use it with $g(x|y)$ for the *fixed* y.

From the above expression for dF, we find

$$F(y) = y^{-n},$$
$$R_1 = y^{-n}, \qquad \text{or} \qquad y = R_1^{-1/n}.$$

This fixes y. Now:

$$\int_0^x g(x'|y)\,dx' = \int_0^x y e^{-x'y}\,dx' = -e^{-x'y}\big|_0^x = 1 - e^{-xy},$$

$$R_2' = 1 - e^{-xy}, \quad \text{or} \quad R_2 = 1 - R_2' = e^{-xy}, \quad \text{or} \quad x = -\frac{1}{y}\log R_2.$$

We can generate this distribution with two pseudorandom numbers and simple functions.

8.3 Acceptance Rejection Method

Another method of considerable use is the acceptance rejection method of J. Von Neumann. It is illustrated in Fig. 8.1.

Consider a distribution that is non-zero between 0 and 1 and whose density function, $f(x)$, is less than the constant f_0 throughout that region. Suppose we wish to generate a set of values, x_i, which have density function f. We generate two pseudorandom numbers R and x, each uniformly distributed between 0 and 1. We reject x if $Rf_0 > f(x)$ and accept it if this inequality is not satisfied. The set of retained x then has the desired density function $f(x)$. Note that the normalizations are adjusted such that f_0 is always greater than $f(x)$. An alternative procedure is to keep all events, but give each event a weight $f(x)/f_0(x)$.

A variant of this method is known as "importance sampling" and is illustrated in Fig. 8.2. Here we modify the acceptance rejection method by finding a function $g(x)$ that is easy to generate in the interval, which roughly matches the density function $f(x)$, and for which a constant C exists such that $f_0(x) = Cg(x) > f(x)$ throughout the interval. x_i are chosen according to $g(x)$ and rejected if $Rf_0(x) > f(x)$, where

Fig. 8.1 Illustration of acceptance rejection method. Two pseudorandom numbers R and x are used. The first is used to pick a value of x. The second is used to accept the event if $f(x)/f_0 > R$ and reject otherwise

Fig. 8.2 Importance sampling. This method is similar to the acceptance rejection method, but the first pseudorandom number x is not picked uniformly, but is picked to follow a distribution approximating the desired one

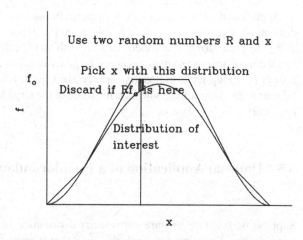

R is again a pseudorandom number uniformly distributed between 0 and 1. This method is more efficient than the first acceptance rejection method in that the test density function more closely matches the desired distribution than does the uniform distribution used in the first method.

8.4 Computer Pseudorandom Number Generators

The computer is an indispensable tool for probability and statistics problems. Several of the exercises below require computer use. It is important to get used to handling this tool.

Random numbers generated by a computer are not really random, but "pseudorandom". One has to be careful that unexpected correlations are not introduced. A great deal of work has been done on this and quite good algorithms now exist.

In general, though, it is important to keep this in mind when doing problems. I know a number of times when these correlations have caused serious problems. Some discussions of generating random numbers and the problems are given in the following references which may not be quite up to date, but give a good discussion of how the problems have been attacked. (James 1988, 1994, 1990; L'Ecuyer 1988; Marsaglia and Zaman 1987; Marsaglia 1985; Lüscher 1994) The pseudorandom number routines that are in standard libraries have often not been of the highest quality, although they will be quite sufficient for the problems in this text. CERN now has several excellent random number routines for a uniform, zero to one, distribution as well as non-uniform ones. For the uniform distribution, there is TRandom(5 ns), TRandom1(105 ns), TRandom2(7 ns), and TRrandom3(10 ns), where the time is the time to generate a single random number. If an array of random numbers is generated, the times are slightly better. In these routines, you can provide one or a few random numbers to serve as starting points. For details, refer to the CERN-root-random-number discussion.

Appendix B shows one method to generate uniformly distributed or normally distributed numbers in C++. These allow you to repeat Monte Carlo calculations using the same string of numbers, or start with an entirely new series.

Of course, true random numbers can be generated. For example, the time between decays of a long-lived radioactive source are truly random, but for generating the billions of random numbers needed for some of the large Monte Carlo programs, not practical.

8.5   Unusual Application of a Pseudorandom Number String

Suppose we want to measure the velocity distribution of a beam of atoms. We can use a shutter, a rotating disk with slots in it that permits the beam to pass, followed by a detector at a distance L that measures the signal as a function of time. If there is one narrow slot, a good resolution is obtained for velocity, but at the cost of poor efficiency.

If the shutter transmission function is $A(t)$, the signal $S(t)$ from the detector is

$$S(t) = K \int_0^\infty A(t - L/v) f(v) dv + B(t),$$

where $f(v)$ is the probability density function for velocity, K is a constant, and $B(t)$ is the background.

Let $T = L/v$ be the time to go from shutter to detector. The probability density function for T is $g(T) = f(v|\partial v/\partial t)$, since $g(T)dT = f(v)dv$. Then

$$S(t) = K \int_0^\infty A(t - T)g(T)dT + B(t).$$

Divide the rim of the disk into N parts, each of which has a slot or no slot. Have the detector integrate the signal over a time corresponding to the passage of one slot length, i.e., the period of rotation divided by N. To average the signal, assume that one measures over M revolutions of the disk.

$$S_n = KM \sum_{m=1}^{N} A_{n-m} g_m + M B_n^c + M^{1/2} B_n^i.$$

Here the background has been divided into a coherent and an incoherent part. Each part is assumed to have a mean of 0. The incoherent part is random from revolution to revolution. Hence, this is a random walk and the measurement will be proportional to $M^{1/2}$. The coherent background is random from slot to slot, but is correlated from revolution to revolution, and is proportional to M.

The crucial idea (Hirschy and Aldridge 1971) is to arrange these slots according to a pseudorandom sequence X_n, where $X_n = \pm 1$. Suppose N is odd and $\sum_{n=1}^{N} X_n = 1$. If X_n is $+1$, there is a slot at location n, and, if $X_n = -1$, there is no slot at location n. Then, $A_n = (1/2)(X_n + 1)$. Note that slightly more than half of the disk is slot, resulting in high efficiency.

Consider the correlation function $C_{AX}(\tau) \equiv \sum_{n=1}^{N} X_n A_{n+\tau}$. Take n as cyclical in this equation, i.e., start over if $n + \tau > N$. If $\tau = 0$,

$$C_{AX}(0) = \sum_{n=1}^{N} X_n A_n = (1/2) \sum_{n=1}^{N} X_n (X_n + 1) = (1/2)(N + 1).$$

If $\tau \neq 0$, to the extent this is a pseudorandom string, $\sum_{n=1}^{N} A_{n+\tau} X_n = N \overline{A}_{n+\tau} \overline{X}_n$. Now $\overline{A}_{n+\tau} = (N + 1)/(2N)$, the fraction of X_n's which are $+1$. Since $\sum_{n=1}^{N} X_n = 1$, $\overline{X}_n = 1/N$. Hence, $\sum_{n=1}^{N} X_n A_{n+\tau} = N \overline{X}_n \overline{A}_{n+\tau} = N \times (1/N) \times (N + 1)/(2N) = (N + 1)/(2N)$.

$$C_{AX}(\tau \neq 0) = (N + 1)/(2N).$$

$$\sum_{n=1}^{N} X_n X_{n+\tau} = \sum_{n=1}^{N} (\overline{X}_n)^2 = N \left(\frac{1}{N} \right)^2 = 1/N \quad (\tau \neq 0).$$

What is $C_{BX} \equiv \sum_{n=1}^{N} X_n B_{n+\tau}$ for any τ?

$$\overline{C}_{BX}(\tau) = \sum_{n=1}^{N} X_n B_{n+\tau} = N \overline{X} \overline{B}_\tau = \overline{B} = 0,$$

since $\overline{B} = 0$.

Cross-correlate the signal $S(t)$ with the pseudorandom string.

$$\sum_{n=1}^{N} X_n S_{n+\tau} = \sum_{n=1}^{N} \left[\left(KM \sum_{m=1}^{N} A_{n-m+\tau} X_n g_m \right) \right.$$
$$\left. + M B_{n+\tau}^c X_n + M^{1/2} B_{n+\tau}^i X_n \right]$$
$$= KM g_\tau (1/2)(N + 1) + \sum_{m \neq \tau} KM g_m (N + 1)/(2N).$$

The mean background is 0 and the unwanted velocity terms are reduced by $1/N$ per point. The sum of the unwanted wrong velocity terms are of the same order as the signal. However, if each time slice is narrow compared to the interval of significant variation, this sum can be considered as a constant background term and subtracted out.

The fluctuations of background and signal do give a contribution. Consider

$$\overline{C^2}_{BX} = (1/N) \sum_{k=1}^{N} C^2_{BX}(k) = (1/N) \sum_{k=1}^{N} \sum_{m=1}^{N} \sum_{n=1}^{N} (B_{n+k} X_n B_{m+k} X_m)$$

$$= (1/N) \sum_{k=1}^{N} \left(\sum_{n=1}^{N} \left[B_{n+k}^2 X_n^2 + \sum_{m \neq n} B_{n+k} B_{m+k} X_m X_n \right] \right).$$

The mean value of the second term is 0 since $\overline{B} = 0$, and X and B are uncorrelated. Note $X_n^2 = +1$.

$$\overline{C^2}_{BX} = (1/N)N \sum_{k=1}^{N} \overline{B^2} = N \overline{B^2}.$$

$$(\overline{C^2}_{BX})^{1/2} \propto \sqrt{N}.$$

Hence, the r.m.s. background is down by $1/\sqrt{N}$. Similarly, the r.m.s. variation of the sum of the unwanted signals is down by $1/\sqrt{N}$ from the signal.

The study of efficient Monte Carlo techniques has grown enormously in recent years. In this chapter, we have tried to introduce some of the basic techniques and, in the homework exercises, we will try to introduce you to the task of using the computer for Monte Carlo simulation. Since Monte Carlo techniques are typically used for applications that are too complicated to work out analytically, they tend, in practice, to often be considerably more difficult than the simple problems here. Some simulations now existing take over 1 MIP-hour to generate one event!

8.6   Worked Problems

WP8.1 Find a method of computing a set of pseudorandom numbers distributed with a density function $f = \frac{1}{2}(1 + \alpha x)$ for $-1 \leq x \leq 1$ with α a constant of magnitude less than 1. You are given a set of pseudorandom numbers, R_i, uniformly distributed between 0 and 1.

Answer:

$$F(x) = \int_{-1}^{x} f(x) = \frac{1}{2} \int_{-1}^{x} (1 + \alpha x') \, dx' = \frac{1}{2} \left(x + 1 + \frac{\alpha}{2}[x^2 - 1] \right)$$

$$= \frac{1}{2} \left(1 - \frac{\alpha}{2} \right) + \frac{x}{2} + \frac{1}{4} \alpha x^2 = R.$$

$$0 = \frac{\alpha x^2}{2} + x + \left(1 - \frac{\alpha}{2} - 2R \right),$$

or

$$x = \frac{-1 \pm \sqrt{1 - 2\alpha \left(1 - \frac{\alpha}{2} - 2R \right)}}{\alpha}.$$

Pick the plus sign above since we want the solution to fall between -1 and 1. Note: This example applies to the decay of the polarized spin $\frac{1}{2}$ resonances with $x = \cos\theta$.

WP8.2 Let

$$x = \sqrt{-2\sigma^2 \log R_1} \cos 2\pi R_2,$$

$$y = \sqrt{-2\sigma^2 \log R_1} \sin 2\pi R_2,$$

where R_1, R_2 are independent pseudorandom numbers uniformly distributed between 0 and 1. Show x and y are independent numbers that are normally distributed with mean 0 and variance σ^2. This is a very convenient relation for generating normally distributed numbers on a computer. This generates normally distributed random numbers a pair at a time. On the first call, two numbers are generated and one returned. On the second call, the second is returned, etc.

Answer: Let

$$r^2 = x^2 + y^2 = -2\sigma^2 \log R_1.$$

We see the distribution function for r^2 is

$$F = 1 - R_1 = 1 - e^{-r^2/2\sigma^2}.$$

Thus, the density function for r^2 is

$$f_1(r^2)\, dr^2 = \frac{1}{2\sigma^2} e^{-r^2/2\sigma^2}\, dr^2.$$

The density function for R_2 is

$$f_2(R_2)\, dR_2 = 1 \times dR_2,$$

$$f(r^2,\ R_2)\, dr^2\, dR_2 = f_1(r^2) f_2(R_2)\, dr^2\, dR_2.$$

Let $\theta = 2\pi R_2$.

$$f(r^2,\ R_2)\, dr^2\, dR_2 = f_1(r^2)\, dr^2\, \frac{d\theta}{2\pi}.$$

But $dr^2\, d\theta = 2r\, dr\, d\theta = 2\, dx\, dy$, where

$$r^2 = x^2 + y^2, \qquad x = r\cos\theta,\ y = r\sin\theta.$$

$$f(x,\ y)\, dx\, dy = f(r^2,\ R_2)\, dr^2\, dR_2 = \frac{1}{2\sigma^2} e^{-r^2/2\sigma^2} \frac{1}{2\pi} 2\, dx\, dy,$$

$$f(x,\ y)\, dx\, dy = \frac{1}{\sqrt{2\pi\sigma^2}} e^{-x^2/2\sigma^2}\, dx\, \frac{1}{\sqrt{2\pi\sigma^2}} e^{-y^2/2\sigma^2} \frac{2}{2}\, dy.$$

Thus, $f(x, y) \, dx \, dy = f_3(x) \, dx \, f_4(y) \, dy$, i.e., breaks up into the product of two separate density functions, each independent of the value of the other parameter. Hence, x and y are independent and are each normally distributed.

8.7 Exercises

8.1 A distribution of events as a function of energy is expected to have a Breit–Wigner resonance (M, Γ) and a background density function proportional to $1/(a + bE)^3$.

It is expected that the number of events in the resonance is one-third of the total number of events. Show how to generate a Monte Carlo distribution of the resultant distribution using random numbers uniformly distributed between 0 and 1.

8.2 Devise a fast way to generate a Poisson distribution using Monte Carlo methods assuming λ is less than 20, using random numbers uniformly distributed between 0 and 1. (Hint: Consider using a table.)

8.3 From this point on, a number of the exercises will ask you to draw graphs and make fits to data. Both kinds of these problems ask you to calculate a set of data, say x vs y and then operate on them. I suggest that it may be simplest to just calculate the set of data in your favorite language and print the results out in a file. That file can be read in by the plotting or minimizing program. In the Appendix I give a short primer on using the CERN root system for these operations, but feel free to generate plots from any plotting system with which you are familiar. Fitting data to theory is a bit more complicated. In the exercises, I assume you are using the CERN program MINUIT for fits.

This first computer exercise is just to acquaint you with the very basic ideas. Generate 1000 numbers randomly distributed between 0 and 1 and plot the results in a histogram. The default random number generator in your code should be adequate.

8.4 In this exercise, we will examine the efficiency of detection of events in an experiment. You will generate 300 events of D^0 decays and see if each event is detectable. To be detectable, the D^0 has to live sufficiently long that the decay point can be seen, it has to decay by the right mode, and the resulting particles from the decay have to go into the detection apparatus.

Suppose you produce D^0 mesons with a lab momentum of 5 GeV/c in the z direction. The D^0 meson has a mass of 1.864 GeV and $c\tau$ = velocity of light × lifetime of 0.0128 cm.

The D^0 decays isotropically into a k^{*-} and a π^+ about 5% of the time. The mass of the k^{*-} is 0.888 GeV. The mass of the π^+ is 0.1396 GeV. For the present exercise, we make the artificial, but simplifying, assumption that the k^{*-} is essentially stable.

(a) Pick two pseudorandom numbers and use them to determine the directions (θ, ϕ) of the resulting k^{*-} and π^+ assuming isotropic D^0 decay in the D^0 center of mass. Hint: Remember that the element of solid angle can be written as

$d\Omega = d(\cos\theta)\,d\phi$.

A useful formula is that for a two-body decay of a particle of mass M at rest into particles of mass m_1 and m_2, we have

$$E_1 = (M^2 - m_2^2 + m_1^2)/2M.$$

For this exercise, it is convenient to use a Lorentz transformation function (LOREN) given below

This will take us from a particle with a 4-momentum vector $B(\vec{p}, E)$ in system Σ to the momentum of the same particle now called X in system Σ'. If a particle of mass M (arbitrary) is at rest in Σ', then it has momentum A in Σ. (This gives us the needed information for β.) For the present purpose, we can take A as the D momentum with spatial components reversed ($-\mathbf{p}_D, E_D$). We are taking Σ' as lab and Σ as rest system of the D. $B = p_x, p_y, p_z, E$ of the particle we want in the Σ reference system (dimension 4):

$X = p_x, p_y, p_z, E$ of particle we want in Σ' system (output of routine)

```
#include <cmath> // allows access to functions like log
// and exp
#include <vector> // needed to use vectors
#include <random> // random number generation
#include <iostream>
#include <fstream>
#include <math.h>
// LOREN
double DOT3(vector<int> a, vector<int> b) {
return a[1] * b[1] + a[2] * b[2] + a[3] * b[3];
}
double loren(double a[4], double b[4], double x[4]) {
double beta[3];
xma = sqrt(pow(a[4], 2) - DOT3(a, a));
gam = a[4] / xma
pp = DOT3(b, a) * gam / ((gam + 1) * a[4]);
for(int i = 1; i < 4; ++i) {
beta[i] = a[i] / a[4];
x[i] = b[i] + gam * beta[i] * (pp - b[4]);
}
x[4] = gam * (b[4] - DOT3(beta, b));
}
```

(b) Next find the efficiency of detection. The D^0 must decay by the mode given (5% probability). Generate 300 events with this decay mode. (There is no point throwing out 95% of the events at this point. It only makes the program run longer. Just put in the decay fraction in the efficiency calculation at the end.) The event is not detected if the D^0 decays before 0.02 cm. (You need one more random number to represent the exponential decay). Assume next that, because of experimental limitations, the event is not detected if either the k^{*-} or π^+ is within 15° of the z axis.

To test the program, make histograms of the decay length, the distribution of angles of the k^- and π^+, and possibly other interesting quantities.

8.5 Let us consider an example of the importance sampling variant of the acceptance rejection method. Consider again the problem of generating normally distributed pseudorandom numbers for $x > 0$.

Start with a function $g(x) = a_1 e^{-a_2 x}$. This is not to be a normalized density function, but a function always larger than the normal density function for $x > 0$.

(a) As the constants a_1, a_2 are decreased, eventually the exponential will touch the normal distribution at first at a single point. Suppose we have that condition. Find the relation this imposes between a_1 and a_2.

(b) Given the above relation, we can still change the exponential to have minimum area. Find the value of a_1 and a_2 to do this. (Remember to select the solution corresponding to the exponential touching the outer edge of the normal distribution, i.e., having more area than the normal distribution not less.)

(c) By taking the ratio of areas of the exponential and normal functions for $x > 0$, find the efficiency of this method, i.e., the fraction of the time the number is accepted.

(d) Program the procedure for doing this and generate 200 normally distributed numbers. Plot the distribution of the exponential and the normally distributed numbers. The CPU time per normal pseudorandom number depends on the number of uniform pseudorandom numbers needed (here two) and the number of special functions used (logs, exponentials, sines, etc.). Try to do this with only two special functions per normal number. Compare the efficiency of your program with the efficiency calculated in (c). Note: When picking the first (exponentially distributed) pseudorandom number, you must use a normalized exponential density function.

This is a good method of generating normally distributed pseudorandom numbers, but not fast as the method outlined in Worked Problem 8.2.

References

Hirschy V (1971) Aldridge. Rev Sci Inst 42:381–383

James F (1988) A review of pseudorandom number generators. Technical Report DD/88/22, CERN-Data Handling Division, CERN, CH-1211, Geneva 2, Switzerland

James F (1990) A review of pseudorandom number generators. Comp Physics Comm 60:329–344

James F (1994) Ranlux: a fortran implementation of the high-quality pseudorandom number generator of lüscher. Comp Phys Comm 79:110

L'Ecuyer P, (1988) Efficient and portable combined random number generators. Comm ACM 31:742

Lüscher M (1994) A portable high-quality random number generator for lattice field theory simulations. Comp Phys Comm 79:100

Marsaglia G (1985) A current view of random number generators. In: Billard L (ed) Computer Science and Statistic: Proceedings of the sixteenth symposium on the interface, Elsevier Science Publishers, North Holland, Amsterdam

Marsaglia G, Zaman A (1987) Toward a universal random number generator. Technical Report
    FSU-SCRU-87-50, Florida State University, Tallahasee, FL 32306-3016
Rubenstein R (1981) Simulation and the Monte Carlo method. Wiley, New York
Yost G (1985) Lectures on probability and statistics. Tech. Rep. LBL-16993, Lawrence Berkeley
    Laboratory Report

Chapter 9
Queueing Theory and Other Probability Questions

Abstract Queueing theory is the theory of standing in lines. We stand in lines waiting for a teller machine at a bank, in checkout lines at grocery stores, and in many other places. The Internet is probably the most complicated present example of the use of queues. If you belong to an organization which maintains several central printers and you submit jobs for printing, there are queues for each printer, or as we will see, a clever single queue. For pulsed beam accelerator experiments, there are queues set up for quick initial processing. It is necessary to know how large the queues need to be and what fraction of events are lost. We will discuss other examples of queueing. We will discuss Markov chains briefly. We will look at birth and death problems, which are fundamental in many fields of science, and for your amusement we will discuss games of chance, and some general theorems under the subject of gambler's ruin.

9.1 Queueing Theory

We start with an example of this process. Suppose in a counting experiment events are incoming randomly with an average rate of λ_I per second. Each event must be processed. Suppose the time for processing is randomly distributed with, at any time, a conditional probability $\lambda_O dt$ of an event finishing and being output, if it has not yet finished. In order to avoid loss of events during processing time, an input buffer is used. If an incoming event would overflow the buffer, it is lost. How large a buffer N is needed in order that only a fraction L of the incoming events will be lost?

There will be initial transients when the apparatus is turned on, but the equilibrium situation is easily calculated. In equilibrium, the probability that k events are in the buffer P_k will be constant in time for each k. Thus, the probability of entering the buffer must be equal to the probability of leaving it. If the number of events in the buffer is k and $0 < k < N$, then

$$P_{k+1}\lambda_O dt + P_{k-1}\lambda_I dt = P_k(\lambda_I dt + \lambda_O dt). \qquad (9.1)$$

© Springer Nature Switzerland AG 2020

B. P. Roe, *Probability and Statistics in the Physical Sciences*,
Undergraduate Texts in Physics,
https://doi.org/10.1007/978-3-030-53694-7_9

For the special situations that $k = 0$ or $k = N$,

$$P_1 \lambda_O dt - P_0 \lambda_I dt = 0; \quad -P_N \lambda_O dt + P_{N-1} \lambda_I dt = 0. \tag{9.2}$$

The solution for these special situations is $P_1/P_0 = P_N/P_{N-1} = \lambda_I/\lambda_O$. For intermediate occupancy try $P_k = CR^k$, where C and R are constants.

$$R^{k+1} \lambda_O + R^{k-1} \lambda_I = R^k (\lambda_O + \lambda_I),$$

$$R^2 \lambda_O + \lambda_I - R(\lambda_O + \lambda_I) = 0.$$

This quadratic equation has as a solution either $R = \lambda_I/\lambda_O$ or $R = 1$. Choose the first solution. Then for any k, $P_k = CR^k$, where $R = \lambda_I/\lambda_O$. C can now be easily determined. The total probability to be in some state is 1. $1 = P_{\le N} = C \sum_{i=0}^{N} R^i$, or $C = 1/ \sum_{i=0}^{N} R^i$. However, $(1 - R) \sum_{i=0}^{N} R^i = 1 - R^{N+1}$ as the intermediate powers cancel term by term. Therefore, $C = (1 - R)/(1 - R^{N+1})$.

$$P_k = ([1 - R]R^k)/(1 - R^{N+1}), \tag{9.3}$$

$$P_{\le k} = \sum_{j=0}^{k} ([1 - R]R^j)/(1 - R^{N+1})$$
$$= (1 - R^{k+1})/(1 - R^{N+1}). \tag{9.4}$$

The loss probability is just P_N, the fraction of time the buffer is filled. If a pulse arrives during that fraction of time, it finds a full buffer and is lost. Hence, $L = P_N = (R^N - R^{N+1})/(1 - R^{N+1})$.

Suppose $N \to \infty$. This corresponds to an unlimited buffer. Then for $R < 1$, $P_k = (1 - R)R^k$; $P_{\le k} = 1 - R^{k+1}$. We leave it as an exercise to show that for an infinite buffer

$$\mathbb{E}\{k\} = \frac{R}{1 - R}; \quad \text{var}\{k\} = \frac{R}{(1 - R)^2}. \tag{9.5}$$

Reinterpret this problem as customers standing in a line. R is called the traffic intensity, and events are customers in the usual queueing notation. The probability that the server is free is $P_0 = (1 - R)$. The mean queue size is $R/(1 - R)$. The probability of more than three customers is R^4. To be efficient, it is desirable to keep the servers busy. However, if the servers are almost always busy, the queue size becomes very long. Table 9.1 indicates some representative values of these quantities. It is seen from this table that it is undesirable to have more than $R = 0.7$ or 0.8 (i.e., 70–80% efficiency for the server) in a single queue because of the rapid increase in line length.

Next, we quote some useful results without proof. Proofs can be found in Feller (1950). How long will a customer have to wait before his service time starts? There is a probability $1 - R$ of no customers being in the queue when he arrives and thus

Table 9.1 Queue Parameters for Various Values of Traffic Intensity. Traffic intensity, the probability the server is free, mean queue size, and the probability of more than three customers in the queue are given

R	$1-R$	$R/(1-R)$	R^4
0.1	0.9	0.111	0.0001
0.3	0.7	0.429	0.0081
0.5	0.5	1.0	0.0625
0.7	0.3	2.333	0.240
0.8	0.2	4.0	0.410
0.9	0.1	9.0	0.656

there being no waiting time. In addition to this discrete point, there is a continuous distribution of waiting times given by

$$p_{\text{start}}(t)dt = R(\lambda_I - \lambda_O)e^{-(\lambda_I-\lambda_O)t}. \tag{9.6}$$

If the service time is included, the distribution of waiting times until the service is completed is continuous and is given by

$$p_{\text{finish}}(t)dt = (\lambda_I - \lambda_O)e^{-(\lambda_I-\lambda_O)t}. \tag{9.7}$$

Consider the problem from the server's point of view. What is the chance of a busy period ever terminating? It can be shown that if $R < 1$, then the probability of termination is 1, while if $R > 1$, the probability is $1/R$. For $R < 1$, the mean length of a busy period is $1/[\lambda_O(1 - R)]$, and the mean number of customers served during a busy period is $1/(1 - R)$.

In Sect. 6.10, we noted that the Weibull distribution gives the probability of the time for first failure of a device at a time between t and $t + dt$ if that probability is a constant independent of the time it has performed properly up to that time.

9.2 Markov Chains

In previous chapters, we have discussed independent (Bernoulli) trials. Markov chains are a first generalization of this. In a Markov chain, the outcome of a trial at time t is not independent independent of history, but depends upon where you are now. It does not depend on how you got to where you are. Define a matrix $P = |p_{ij}|$, the stochastic matrix, where p_{ij} is the probability of going from state i at time t into state j at time $t + 1$. For Markov chains, values in P_{t+1} depend only on values in P_t. Markov Chains are used to numerically estimate uncertainties in the parameters of a model using a sequence of random samples. A detailed discussion of this method is given in Spreagle (2019). The buffer problem had changes in $P(t)_k$ dependent on

$P(t-1)_k$, $P(t-1)_{k-1}$, and hence is a Markov chain. (See Eq. 9.1). In this problem the, stochastic matrix has entries on the main diagonal and the diagonals above and below. Suppose $\lambda_I = 0$. Then only the main diagonal and the diagonal above it are filled. Suppose further that the λ_O are dependent on the state k. This might correspond to a radioactive decay chain, where there are a series of isotopes $A, B, C, \ldots$, where $A \to B \to C \ldots$. For a small interval of time, p_{ij}, $i \neq j$ connects only adjacent elements along the decay chain and is $\lambda_i \Delta t$, where λ_i is the lifetime of the ith element in the chain.

Suppose that the diagonals above and below the main diagonal are filled. This is again a similar problem to the buffer problem, except that now we let the λ's depend on the state. Let $P_k(t)$ refer to the non-equilibrium situation.

$$\frac{dP_k}{dt} = -\lambda_{I,k} P_k - \lambda_{O,k} P_k + \lambda_{I,k-1} P_{k-1} + \lambda_{O,k+1} P_{k+1} \text{ for } k > 0,$$
$$\frac{dP_0}{dt} = -\lambda_{I,0} P_0 + \lambda_{O,1} P_1. \tag{9.8}$$

This corresponds to a birth and death problem in population problems. Consider a population, with the birthrate being $k\lambda_I$ and the death rate $k\lambda_O$. This is the linear growth problem, with the probability of a birth or death proportional to the size of the population. The linear problem also can be viewed as a one-dimensional random walk problem with the probability of a step to the right being $p = \lambda_I/(\lambda_I + \lambda_O)$ and the probability of a step to the left being $q = \lambda_O/(\lambda_I + \lambda_O)$. If the equations are viewed as a game, with probability p of winning a round and q of losing a round, then the point $k = 0$ corresponds to the gambler's ruin.

The probability of ultimate extinction is 1 if $p < q$ and is $(q/p)^r$ if $p > q$ and the initial state of the system is one with r events present. In terms of λ_I and λ_O, the probability of ultimate extinction is $(\lambda_O/\lambda_I)^r$ if $\lambda_I > \lambda_O$. It is interesting that there is a finite chance that the population is never extinguished. Basically, this occurs since almost all of the loss occurs for small population numbers. If the population grows past a certain point, it is reasonably safe.

The general problem with $\lambda_{I,k}$, $\lambda_{O,k}$ not necessarily linear, can be viewed as a generalization of the queueing problem, with the arrival and serving times dependent on the length of line. It is left as an exercise to show that the equilibrium probabilities for this situation are given by

$$P_k = \frac{\prod_{i=0}^{k-1} \lambda_{I,i}}{\prod_{j=1}^{k} \lambda_{O,j}} P_0, \quad k \geq 1. \tag{9.9}$$

P_0 is chosen for correct normalization

$$P_0 \equiv \frac{1}{s} = \frac{1}{1 + \sum_{k=1}^{\infty} \left(\prod_{i=0}^{k-1} \lambda_{I,i} / \prod_{j=1}^{k} \lambda_{O,j} \right)}. \tag{9.10}$$

A new definition for traffic intensity is needed for this example. Define R to be the reciprocal of the radius of convergence of the power series

$$\sum_{k=1}^{\infty} \frac{z^k \prod_{i=0}^{k-1} \lambda_{I,i}}{\prod_{j=1}^{k} \lambda_{O,j}}.$$

Clearly, if s is infinite, then $R \geq 1$ and we are getting too many customers. Let us consider several examples.

(1) Suppose a long line discourages customers and set $\lambda_{O,k} = \lambda_O$; $\lambda_{I,k} = \lambda_I/(k+1)$. The solution then becomes

$$P_k = e^{-\lambda_I/\lambda_O} \frac{(\lambda_I/\lambda_O)^k}{k!} \quad k > 0. \tag{9.11}$$

(2) Assume there are always enough servers. $\lambda_{I,k} = \lambda_I$; $\lambda_{O,k} = k\lambda_O$. We again obtain Eq. 9.11

(3) Assume there are m servers. This is the "quickline" concept, one line and several servers, often used in bank and airport lines. $\lambda_{I,k} = \lambda_I$, $\lambda_{O,j} = j\lambda_O$, $j \leq m$, $\lambda_{O,j} = m\lambda_O$, $j > m$. Here, let $R = \lambda_I/(m\lambda_O)$. Then

$$P_k = \frac{(mR)^k}{k!} P_0, \quad k < m$$

$$= \frac{m^m R^k}{m!} P_0, \quad k \geq m \tag{9.12}$$

$$\frac{1}{P_0} = 1 + \left(\sum_{i=1}^{m-1} \frac{(mR)^i}{i!} \right) + \frac{R^m m^m}{m!(1-R)}.$$

How well does this "quickline" concept work compared to several single server queues. Consider $m = 2$. Here $P_0 = (1-R)/(1+R)$; $E\{k\} = 2R/(1-R^2)$. For a single server with half the input rate, the mean length in Eq. 9.5 was $R/(1-R)$, with R the same value as for the present problem. Furthermore, since we have two servers here, we should compare 1/2 of the "quickline" $E\{k\}$, i.e., $R/(1-R^2)$ to $R/(1-R)$. For R near 1, the mean line length per server and hence the waiting times are appreciably shorter with a "quickline." The occasional long serving time does not tie up the whole queue with a "quickline".

(4) Next suppose we consider a telephone exchange with n trunklines. This corresponds to having n servers and having waiting room for only n customers. $\lambda_{I,k} = \lambda_I$ for $k < n$ and $\lambda_{I,k} = 0$ for $k \geq n$. $\lambda_{O,k} = k\lambda_O$ for $k \leq n$. Then

$$P_k = \frac{(\lambda_I/\lambda_O)^k}{k!(\sum_{i=o}^{n}(\lambda_I/\lambda_O)^i/i!)}, \quad k \leq n. \tag{9.13}$$

(5) Suppose m repair people are responsible for n machines in a factory. If k of the machines are already broken down, there are fewer left which might break down and add to the queue. If $k \le m$, then the broken machines are all being serviced. If $k > m$, then $k - m$ broken machines are waiting for a serviceman to be available. $\lambda_{I,k} = (n - k)\lambda_I$, $k \le n$, and $\lambda_{I,k} = 0$, $k > n$. $\lambda_{O,0} = 0$; $\lambda_{O,k} = k\lambda_O$, $k \le m$ and $\lambda_{O,k} = m\lambda_O$, $k > m$. Let $R' = \lambda_I/\lambda_O$. R' is called the servicing factor.

$$P_k = \frac{n!(R')^k}{k!(n-k)!}P_0 \text{ for } k \le m,$$

$$= \frac{n!(R')^k}{m^{k-m}m!(n-k)!}P_0 \text{ for } k > m, \tag{9.14}$$

$$\frac{1}{P_0} = \sum_{i=0}^{m}\frac{n!(R')^i}{(n-i)!i!} + \sum_{i=m+1}^{n}\frac{n!(R')^i}{m^{i-m}m!(n-i)!}.$$

We leave the proof to the exercises. For one serviceman ($m = 1$)

$$\mathbb{E}\{k\} = n - \frac{\lambda_I + \lambda_O}{\lambda_I}(1 - P_0). \tag{9.15}$$

If we have m repairmen, then the fraction of time that the repair people are busy (the operator utilization) is $u = \sum_{r=0}^{m} r P_r/m + \sum_{r=m+1}^{n} P_r$. The rate of production per machine is $um/(nR')$.

All of the above examples have assumed random arrivals and exponentially distributed service times. This is surprisingly often a good approximation, but it is arbitrary. Questions of rush hours, non-random arrival, or departure time, although not treated here must often be dealt with in practice, as must problems involving series of queues or queues with priorities (computer queues).

9.3   Games of Chance

Suppose one has a series of Bernoulli trials with probability of success p and failure $q = 1 - p$. In Exercise 5.1, we found that the negative binomial distribution gave the probability of having the mth success occur on the rth Bernoulli trial as

$$P(r, m) = \frac{(r-1)!}{(m-1)!(r-m)!}p^m q^{r-m}. \tag{9.16}$$

The mean and variance for the number of failures preceding the mth success is

$$\text{mean} = \frac{mq}{p}, \quad \text{variance} = \frac{mq}{p^2}. \tag{9.17}$$

We leave the proof to the exercises.

If one asks to have a run of r consecutive successes in a string of Bernoulli trials, then the mean number of trials until this string has occurred and its variance are

$$\text{mean} = \frac{1 - p^r}{qp^r}, \quad \text{variance} = \frac{1}{(qp^r)^2} - \frac{2r + 1}{qp^r} - \frac{p}{q^2}. \tag{9.18}$$

Consider now a coin toss game, Larry versus Don. To eliminate trivial complication, we will say that if the game is tied and Larry had led immediately preceding the tie, that Larry still leads. Several theorems can then be obtained which we quote without proof (Feller 1950).

Theorem 9.1 *The probability that Larry leads in 2r out of 2n trials is*

$$P(2r, 2n) = \binom{2r}{r}\binom{2n - 2r}{n - r}2^{-2n}. \tag{9.19}$$

Theorem 9.2 *A rather surprising limit theorem exists. If n is large and $z = r/n$ is the fraction of time during which Larry leads, then the probability that $z < t$ approaches*

$$F(t) = \frac{2}{\pi}\arcsin t^{1/2}. \tag{9.20}$$

In Fig. 9.1, the distribution function $F(t)$ and the density function $f(t)$ are plotted. The density function peaks strongly for t near 0 or 1. The less fortunate player is seldom ahead. If a coin is tossed once per second for a year, there is better than a 10% chance that the less fortunate player leads for less than 2.25 days.

Theorem 9.3 *Let y_i be the duration of the ith game from the beginning until the first time both players are again equal. Let*

$$\bar{y}_k = \frac{1}{k}\sum_{i=1}^{k} y_i. \tag{9.21}$$

$k\bar{y}_k$ *is the number of trials until Larry and Don are tied for the kth time. Now another curious limit theorem exists. For fixed α,*

$$Pr\{\bar{y}_k \leq k\alpha\} \to 2(1 - F(\alpha^{-1/2})) \text{ as } n \to \infty, \tag{9.22}$$

where $F(x)$ is the normal distribution function (integral probability function) with unit variance. This implies that $\mathbb{E}\{\sum_{i=1}^{k} y_i\}$ is proportional to k^2. Thus, the average is not stable, but increases with k. $\mathbb{E}\{\bar{y}_k\}$ is proportional to k.

This occurs because the probability doesn't fall off sufficiently fast for large y. This illustrates that depending on the kind of distribution we have, it can be very dangerous to throw away large deviations in physical measurements. An arbitrary criterion sometimes used for rejecting bad measurements is Chauvenet's criterion.

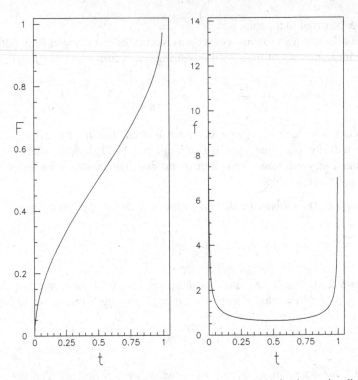

Fig. 9.1 The distribution function $F(t)$ and the density function $f(t)$ for the arcsine distribution $F(t) = \frac{2}{\pi}\arcsin t^{1/2}$

Suppose the deviation of the bad point is Δ and that the probability of a deviation at least as large as Δ is believed to be less than $1/2n$, where n is the number of data points. Chauvenet's criterion is to then reject this point. From the above example, it seems clear that any uniformly applied automatic criterion may cause severe bias in particular cases. Each experiment should be considered on its own.

Theorem 9.4 *The above theorems have implicitly assumed a fair case, with $p = 1/2$ for success. Suppose the game is not fair and p is the probability of Larry's success on a given trial, $q = 1 - p$. Then the probability that equilibrium (0 gain) ever returns is*

$$f = 1 - |p - q|. \tag{9.23}$$

9.4   Gambler's Ruin

Imagine a game of chance (Bernoulli trials) with unit stakes played with each player having a finite capital. The game is to be continued until one person runs out of capital, i.e., is ruined. This is the most dramatic way of phrasing the problem. However, it also enters into physics problems. For instance, we might consider the one-dimensional random walk of a neutron in a material in which it leaves the material if it gets past the edges or, alternatively, it gets absorbed at the edges.

Suppose the initial capital of the player is z and that of his opponent is $a - z$. Play continues until someone is ruined. (This is equivalent to a problem in which a player plays an infinitely wealthy opponent and decides in advance to play until the player is ruined or until the player's capital is increased to a.) Let q_z be the probability of ruin. $1 - q_z$ is then the probability of winning. Let p be the probability of the player winning 1 trial and $q = 1 - p$. Then by examining the situation after one trial, it is easily seen that

$$q_z = pq_{z+1} + qq_{z-1}. \tag{9.24}$$

If we let $q_0 = 1$ and $q_a = 0$, then Eq. 9.24 is true for $1 \leq z \leq a - 1$. It can be easily checked by direct substitution that the solution to this equation is

$$q_z = \frac{(q/p)^a - (q/p)^z}{(q/p)^a - 1}. \tag{9.25}$$

If the game is fair, $p = q = 1/2$, then by using L'Hôpital's rule, one obtains

$$q_z = 1 - \frac{z}{a}. \tag{9.26}$$

Note that this result is independent of the size of the bet on each trial, since it depends only on the ratio of z to a. The limit $a \to \infty$ is the probability of ruin if the opponent is infinitely rich. For a fair game, it is seen from Eq. 9.26 that the probability of ruin is 1. For the general case, what is the effect of changing the stakes for each trial? If the stakes are halved, it is the equivalent of letting z and a be replaced by twice their respective values. The new probability of ruin is then

$$q_z^* = \frac{(q/p)^{2a} - (q/p)^{2z}}{(q/p)^{2a} - 1} = q_z \frac{(q/p)^a + (q/p)^z}{(q/p)^a + 1}.$$

If $q > p$, the coefficient of q_z is greater than unity. Hence, if the stakes get smaller and the odds are unfair, the probability of ruin increases for the player for whom $p < 0.5$. This is easily understood, since, in the limit of very small stakes, the statistics are smoothed out and one must be ruined if the odds are unfair. If the game is decided in one trial, the odds are p to q if the two players start with an equal capital. These are the best odds the lower probability player can get. If you are forced to play an unfair game with the odds against you, bet high!

What is the duration of the game expected to be? If a is finite, then the expected number of rounds in the game can be shown to be

$$\mathbb{E}\{n\} = \frac{z}{q - p} - \frac{a}{q - p} \left(\frac{1 - (q/p)^z}{1 - (q/p)^a} \right). \tag{9.27}$$

In the limit $p = q$, this becomes $\mathbb{E}\{n\} = z(a - z)$. This is infinite in the limit $a \to \infty$. Thus the recurrence time for equilibrium in coin tossing is infinite and the mean time for the first return to any position is infinite.

9.5   Exercises

9.1 Show that for the infinite buffer problem at the beginning of this chapter, $E\{k\} = R/(1 - R)$ and var$\{k\} = R/(1 - R)^2$.

9.2 Show that the equilibrium equations, Eqs. 9.9 and 9.10 satisfy the equilibrium versions of Eq. 9.8.

9.3 Derive Eq. 9.14. Hint: Consider the equilibrium situation with k machines needing service for three cases, $k = 0$, $1 \le k \le m$, and $k > m$. Show these equilibrium relations are satisfied by

$$(k + 1)\lambda_O P_{k+1} = (n - k)\lambda_I P_k, \ k \le m; \qquad (9.28)$$
$$m\lambda_O P_{k+1} = (n - k)\lambda_I P_k, \ k > m.$$

Then show these equations lead to Eq. 9.14.

9.4 For Sect. 9.1, Item 5, repair people servicing machines, calculate for $m = 3$, $n = 20$, $R' = 0.1$ a table for $k = 1 - 10$, showing the values for the number of machines being serviced, waiting to be serviced, repair people idle, and P_k.

9.5 Derive Eq. 9.17 for the mean and variance for the number of failures preceding the mth success.

References

Feller W (1950) Probability theory and its applications, vol 1. Wiley, New York
Spreagle J (2019) A conceptual introduction to markov chain monte carlo methods. http://arxiv. org/org/pdf/1909.123134

Chapter 10
Two-Dimensional and Multidimensional Distributions

Abstract Until now, we have considered mostly problems involving a single random variable. However, often we have two or more variables. We may have an event characterized by energy and angle, temperature and pressure, etc. Sometimes the two variables are completely independent, but often they are strongly correlated. In this chapter, we will examine general two- and n-dimensional probability distributions and also the generalization of the normal distribution to two and more dimensions. We will examine correlation coefficients, ellipses (ellipsoids) of concentration, moment matrices, and characteristic functions.

10.1 Two-Dimensional Distributions

We recall from Chap. 2 that two random variables x_1 and x_2 are independent if and only if their joint density function $f(x_1, x_2)$ factors, i.e., if $f(x_1, x_2) = f_1(x_1)f_2(x_2)$. In general, for two-dimensional distributions, this is not true. We let

$$m_1 \equiv \mathbb{E}\{x_1\}, \qquad m_2 \equiv \mathbb{E}\{x_2\}, \tag{10.1}$$

$$\lambda_{11} \equiv \sigma_1^2 \equiv \mathbb{E}\{(x_1 - m_1)^2\}, \qquad \lambda_{22} \equiv \sigma_2^2 \equiv \mathbb{E}\{(x_2 - m_2)^2\}, \tag{10.2}$$

$$\mathrm{cov}(x_1, x_2) \equiv \lambda_{12} \equiv \lambda_{21} \equiv \mathbb{E}\{(x_1 - m_1)(x_2 - m_2)\}, \tag{10.3}$$

$$\rho \equiv \frac{\lambda_{12}}{\sigma_1 \sigma_2}. \tag{10.4}$$

$\mathrm{cov}(x_1, x_2)$ is called the covariance of x_1 and x_2. ρ is called the correlation coefficient. The matrix Λ of the λ_{ij} is called the moment matrix.

It can be shown that the rank of the matrix Λ is 0 if and only if the entire distribution is non-zero only at a single point. The rank is 1 if and only if the distribution is non-

© Springer Nature Switzerland AG 2020
B. P. Roe, *Probability and Statistics in the Physical Sciences*,
Undergraduate Texts in Physics,
https://doi.org/10.1007/978-3-030-53694-7_10

zero only along a certain straight line, $x_2 = ax_1 + b$, but not at a single point only. The rank is 2 if and only if there is no straight line containing the total mass of the distribution.

Consider

$$Q(t, u) = \mathbb{E}\{[t(x_1 - m_1) + u(x_2 - m_2)]^2\} = \lambda_{11}t^2 + 2\lambda_{12}tu + \lambda_{22}u^2. \quad (10.5)$$

This is the expectation value of a squared quantity and must always be greater than or equal to zero for any t, u. Q is called a non-negative (semi-positive definite) quadratic form. Since Q is non-negative, the discriminant is less than or equal to zero. Thus,

$$\lambda_{11}\lambda_{22} - \lambda_{12}^2 \geq 0. \quad (10.6)$$

This can be seen most easily by setting u or t equal to 1 and looking at the discriminant in the other variable.

If x_1 and x_2 are independent, then the covariance of x_1 and x_2, λ_{12}, is zero as is ρ. If $\rho = 0$, we say the variables are uncorrelated. This does not mean necessarily that they are independent. If ρ is not zero, it can be shown that x_1 and x_2 can always be expressed as a linear function of uncorrelated variables. ρ will always be between -1 and $+1$.

We will illustrate these concepts by using our old friend, the multiple scattering distribution that we considered in Chap. 3. We will take as variables

$$x_1 = \theta, \qquad x_2 = y.$$

Recall that for two-dimensional multiple scattering, we have found the following:

$$m_1 = \overline{\theta} = 0, \qquad m_2 = \overline{y} = 0,$$

$$\lambda_{11} = \sigma_{x_1}^2 = \sigma_\theta^2, \qquad \lambda_{22} = \sigma_{x_2}^2 = \frac{L^2}{3}\sigma_\theta^2, \qquad \lambda_{12} = \overline{y\theta} = \frac{L}{2}\sigma_\theta^2.$$

From this, we can easily calculate the correlation coefficient ρ to be

$$\rho \equiv \frac{\lambda_{12}}{\sigma_1\sigma_2} = \frac{(L/2)\sigma_\theta^2}{\sqrt{\sigma_\theta^2(L^2/3)\sigma_\theta^2}} = \frac{\sqrt{3}}{2}.$$

As we expect, this is a number between -1 and 1 and all dimensional quantities have canceled out of it. We next examine the quadratic form, Q.

$$Q = \sigma_\theta^2\left(t^2 + 2\frac{L}{2}tu + \frac{L^2}{3}u^2\right) = \sigma_\theta^2\left(t^2 + Ltu + \frac{L^2}{3}u^2\right).$$

The discriminant condition becomes

$$\lambda_{11}\lambda_{22} - \lambda_{12}^2 = (\sigma_\theta^2)^2 \left[\left(\frac{L^2}{3} \right) - \left(\frac{L}{2} \right)^2 \right] = (\sigma_\theta^2)^2 \frac{L^2}{12} > 0.$$

It is clear that this condition is satisfied. Let us return to the general treatment.

We are often interested in the best estimate of x_2, given the value of x_1. Thus, consider the conditional probability for x_2 with x_1 fixed. We will let the mean value of x_2 for fixed x_1 be $m_2(x_1)$. The line $x_2 = m_2(x_1)$ is called the regression line of x_2 on x_1. In a similar manner, the line $x_1 = m_1(x_2)$ is called the regression line of x_1 on x_2. If the line obtained is straight, we say this is a case of linear regression.

Suppose that, whether the line is straight or not, we make a linear approximation to it. We set $m_2'(x_1) = a + bx_1$ and try to minimize $E\{(x_2 - a - bx_1)^2\}$. This corresponds to minimizing the mean square vertical (x_2) distance between points on the distribution and a straight line. We obtain the line given by

$$\frac{x_2 - m_2}{\sigma_2} = \rho \left(\frac{x_1 - m_1}{\sigma_1} \right) \qquad (x_2 \text{ on } x_1). \tag{10.7}$$

This is a best fit for a linear regression line of x_2 on x_1. Similarly,

$$\frac{x_2 - m_2}{\sigma_2} = \frac{1}{\rho} \left(\frac{x_1 - m_1}{\sigma_1} \right) \qquad (x_1 \text{ on } x_2) \tag{10.8}$$

is the best fit for a linear regression line of x_1 on x_2. (See Figs. 10.1 and 10.2.) Here m_1 and m_2 are the overall means defined at the beginning of this chapter.

The residual variances are

$$\mathbb{E}\{(x_2 - m_2'(x_1))^2\} = \sigma_2^2(1 - \rho^2), \tag{10.9}$$

Fig. 10.1 Regression line of x_2 on x_1 and the linear approximation to this line

Linear approximation

Regression line of x_2 on x_1

x_2

x_1

Fig. 10.2 The linear regression line of x_1 on x_2 and that of x_2 on x_1

$$\mathbb{E}\{(x_1 - m_1'(x_2))^2\} = \sigma_1^2(1 - \rho^2). \tag{10.10}$$

Let us again look at the multiple scattering example. The linear regression line of x_2 on x_1 is

$$\frac{y}{\sqrt{(L^2/3)\sigma_\theta^2}} = \frac{\sqrt{3}}{2}\frac{\theta}{\sqrt{\sigma_\theta^2}},$$

$$y = \frac{L}{2}\theta, \qquad y \text{ on } \theta.$$

Similarly, the linear regression line of x_1 on x_2 is

$$y = \frac{2L}{3}\theta, \qquad \theta \text{ on } y.$$

The residual variance for x_2 on x_1 is given by

$$\mathbb{E}\{(x_2 - m_2'(x_1))^2\} = \mathbb{E}\left\{\left(y - \frac{L}{2}\theta\right)^2\right\} = \sigma_2^2(1 - \rho^2)$$

$$= \frac{L^2}{3}\sigma_\theta^2\left(1 - \frac{3}{4}\right) = \left(\frac{L^2}{12}\right)\sigma_\theta^2.$$

Similarly, the residual variance for x_1 on x_2 is given by

$$\mathbb{E}\left\{\left(\theta - \frac{3}{2L}y\right)^2\right\} = \tfrac{1}{4}\sigma_\theta^2.$$

Fig. 10.3 Illustration of a uniform distribution having mean m and standard deviation σ

For the multiple scattering case, these relations can easily be shown directly. For example, the first is

$$\mathbb{E}\left\{y^2 - L\theta y + \frac{L^2}{4}\theta^2\right\} = \frac{L^2}{3}\sigma_\theta^2 - \frac{L}{2}L\sigma_\theta^2 + \frac{L^2}{4}\sigma_\theta^2 = \frac{L^2}{12}\sigma_\theta^2.$$

We note that the residual variances are indeed considerably smaller than the overall variances since they have the best estimate of this parameter using the other, correlated, parameter subtracted out. A very strong correlation implies a very small residual variance, and finally for the limiting case of $\rho = 1$, the residual variance will be zero.

A useful tool to help visualize the spread of a two-dimensional distribution is the ellipse of concentration. In one dimension, we can find a uniform distribution with the same mean and variance as the distribution of interest. A uniform distribution from $m - \sigma\sqrt{3}$ to $m + \sigma\sqrt{3}$ (see Fig. 10.3) has mean m and standard deviation σ. In between these limits, the value $f(x) = 1/(2\sigma\sqrt{3})$ is determined by normalization.

In two dimensions, we can find an ellipse with a uniform distribution inside and zero outside such that m_1, m_2, σ_1, σ_2, ρ are the same as for the distribution of interest. In fact, we can show that the ellipse

$$\frac{1}{1-\rho^2}\left(\frac{(x_1 - m_1)^2}{\sigma_1^2} - \frac{2\rho(x_1 - m_1)(x_2 - m_2)}{\sigma_1\sigma_2} + \frac{(x_2 - m_2)^2}{\sigma_2^2}\right) = 4 \quad (10.11)$$

is the appropriate ellipse. It is called the ellipse of concentration. Distribution A is said to have a greater concentration than distribution B if the ellipse of concentration of A is wholly contained in the ellipse of concentration of B. This ellipse is a measure of the spread of the distribution.

So far, we have been discussing a general two-dimensional distribution. Let us now specialize to a two-dimensional normal distribution. This distribution is defined by the density function:

$$f(x_1, x_2) \equiv \frac{1}{2\pi \sigma_1 \sigma_2 \sqrt{1 - \rho^2}} e^{-Q^{-1}(x_1 - m_1, x_2 - m_2)/2}, \tag{10.12}$$

where

$$Q^{-1}(s, r) \equiv \frac{1}{1 - \rho^2}\left(\frac{s^2}{\sigma_1^2} - \frac{2\rho sr}{\sigma_1 \sigma_2} + \frac{r^2}{\sigma_2^2}\right) = (r \;\; s)\, \Lambda^{-1}\binom{r}{s}. \tag{10.13}$$

$$Q(t, u) = \lambda_{11}t^2 + 2\lambda_{12}tu + \lambda_{22}u^2 = (t \;\; u)\, \Lambda\binom{t}{u}. \tag{10.14}$$

Note that the term in the exponent in Eq. 10.12 corresponds to the left-hand side of the ellipse-of-concentration expression.

The characteristic function for this distribution is

$$\phi(t, u) \equiv \mathbb{E}\{e^{i(tx_1 + ux_2)}\} = e^{i(m_1 t + m_2 u) - Q(t,u)/2}. \tag{10.15}$$

For this distribution, the conditional distribution of x_1 with x_2 fixed is a one-dimensional normal distribution. The regression lines for the two-dimensional normal distribution are exactly linear and are given by the equations above. These regression lines can be shown to fall at the center of the appropriate conditional distribution and to intersect the ellipse of concentration where the tangents are vertical and horizontal. These geometrical relationships are shown in Fig. 10.4.

Again, let us use our example of two-dimensional multiple scattering, now assuming that, at least approximately, the distribution is a normal one. The coefficient in front of the exponential becomes

Fig. 10.4 Regression lines and the ellipse of concentration for a two-dimensional normal distribution

$$\frac{1}{2\pi\sigma_1\sigma_2\sqrt{1-\rho^2}} = \frac{1}{2\pi\sigma_\theta^2(L/\sqrt{3})(1/2)} = \frac{\sqrt{3}}{\pi\sigma_\theta^2 L}.$$

The term in the exponential becomes

$$Q^{-1}(x_1, x_2) = \frac{4}{\sigma_\theta^2}\left[\theta^2 - \frac{2(\sqrt{3}/2)\theta y}{(L/\sqrt{3})} + \frac{y^2}{(L^2/3)}\right] = \frac{4}{\sigma_\theta^2}\left[\theta^2 - \frac{3\theta y}{L} + \frac{3y^2}{L^2}\right].$$

As we noted, this is also the appropriate left-hand side term for the ellipse-of-concentration expression.

Finally, we can easily calculate the characteristic function for the multiple scattering distribution to be

$$\phi(t, u) = e^{-Q/2} = e^{-\sigma_\theta^2(t^2 + Ltu + (L^2/3)u^2)}.$$

10.2 Multidimensional Distributions

For n-dimensional distributions, we proceed in analogy with the two-dimensional case. Let

$$m_i \equiv \mathbb{E}\{x_i\}, \tag{10.16}$$

$$\sigma_i^2 \equiv \mathbb{E}\{(x_i - m_i)^2\}, \tag{10.17}$$

$$\lambda_{ik} \equiv \mathbb{E}\{(x_i - m_i)(x_k - m_k)\}, \tag{10.18}$$

$$\rho_{ik} \equiv \frac{\lambda_{ik}}{\sigma_i\sigma_k}, \tag{10.19}$$

$$\Lambda \equiv (\lambda_{ik}) \equiv \text{moment matrix}, \tag{10.20}$$

$$P \equiv (\rho_{ik}) \equiv \text{correlation matrix}, \tag{10.21}$$

$$Q(t_1, t_2, \ldots, t_n) \equiv \mathbb{E}\left\{\left(\sum_{j=1}^{n} t_j(x_j - m_j)\right)^2\right\} = \sum_{i,k=1}^{n} \lambda_{ik}t_it_k = t^T\Lambda t \tag{10.22}$$

(T = transpose). We note that $\lambda_{ii} = \sigma_i^2$ and $\rho_{ii} = 1$ trivially from the above definitions. Again, Q is a non-negative quadratic form. A theorem on rank analogous to the two-dimensional case can be obtained. The rank of Λ is the minimum number of independent variables. Thus, if the rank is less than n, some variables can be eliminated.

The ellipsoid of concentration is analogously defined. It is that ellipsoid in n-dimensional space such that a uniform distribution with that ellipsoid as a boundary has the same Λ (or P). It is given by

$$g(x_1, x_2, \ldots, x_n) \equiv \sum_{i,k=1}^{n} (\Lambda^{-1})_{ik}(x_i - m_i)(x_k - m_k) = n + 2. \qquad (10.23)$$

Define Λ^{ik} to be $(-1)^{i+k}$ times the determinant of the Λ matrix with row i and column k removed. Λ^{ik}, the cofactor, is to be distinguished from λ_{ik}, which is the i, k element of the matrix Λ. We will similarly denote P^{ik} as the i, k cofactor of P. $|\Lambda|$ will denote the determinant of Λ. Thus,

$$(\Lambda^{-1})_{ik} \equiv \frac{\Lambda^{ki}}{|\Lambda|}. \qquad (10.24)$$

We next look at the analogy of regression lines for the multidimensional case. For simplicity, we assume all m_i are zero. This just involves a translation of coordinates. To find the best linear regression plane of x_i on $x_1, x_2, \ldots, x_{i-1}, x_{i+1}, x_{i+2}, \ldots, x_n$, let

$$\eta_i \equiv i\text{th residual} = x_i - x_i^*, \qquad (10.25)$$

$$x_i^* \equiv \sum_{k \neq i} \beta_{ik} x_k. \qquad (10.26)$$

Here the β_{ik} are constants to be determined, so that $\mathbb{E}\{(x_i - x_i^*)^2\}$ is a minimum. We can then show

$$\beta_{ik} = \frac{-\Lambda^{ik}}{\Lambda^{ii}} = \frac{-\sigma_i P^{ik}}{\sigma_k P^{ii}}. \qquad (10.27)$$

We find that, in contrast to the two-dimensional case, we can now define several different kinds of correlation coefficients. ρ_{ik}, defined above, is called the total correlation coefficient. Another kind of correlation coefficient is the partial correlation coefficient, ρ_{ik}^P. This coefficient measures the correlation of variables i and k with the correlations due to other variables removed. Specifically, we ask for the correlation of x_i and x_k after subtracting off the best linear estimates of x_i and x_k in terms of the remaining variables. Let β_{ik}^j be the value of β_{ik} if variable j is ignored (i.e., integrated over). Thus, let Λ^j be the matrix Λ with the jth row and column removed. Then

$$\beta_{ik}^j = \frac{-(\Lambda^j)^{ik}}{(\Lambda^j)^{ii}}, \qquad (10.28)$$

$$x_i^{*j} \equiv \sum_{k \neq i,j} \beta_{ik}^j x_k, \qquad (10.29)$$

$$\eta_i^j \equiv x_i - x_i^{*j}, \tag{10.30}$$

$$\frac{\mathbb{E}\{\eta_k^i \eta_i^k\}}{\sqrt{\mathbb{E}\{(\eta_k^i)^2\}\mathbb{E}\{(\eta_i^k)^2\}}} = \frac{-P^{ik}}{\sqrt{P^{ii}P^{kk}}} = \frac{-\Lambda^{ik}}{\sqrt{\Lambda^{ii}\Lambda^{kk}}}. \tag{10.31}$$

Still another correlation coefficient is the multiple correlation coefficient ρ_i^M. This coefficient measures the correlation of x_i with the totality of the rest of the variables. We can show that of all linear combinations of the rest of the variables, x_i^* has the largest correlation with x_i. Thus, we define

$$\rho_i^M \equiv \frac{\mathbb{E}\{x_i x_i^*\}}{\sqrt{\mathbb{E}\{x_i^2\}E\{x_i^{*2}\}}} = \sqrt{1 - \frac{|P|}{P^{ii}}}. \tag{10.32}$$

All of these correlation coefficients (ρ_{ik}, ρ_{ik}^P, ρ_i^M) have magnitudes less than or equal to one. In addition, ρ_i^M is greater than or equal to zero.

Next we turn to the multidimensional normal distribution. To simplify notation, we assume that all m_i are zero. To relax this assumption, simply replace x_i by $x_i - m_i$ in the equations given below. We let the density function be $f(x_1, x_2, \ldots, x_n)$. The multidimensional normal distribution is defined by

$$
\begin{aligned}
f(x_1, x_2, \ldots, x_n) &\equiv \frac{1}{(2\pi)^{n/2}\sqrt{|\Lambda|}} \exp\left(-\frac{1}{2}x^T \Lambda^{-1} x\right) \\
&= \frac{1}{(2\pi)^{n/2}\sqrt{|\Lambda|}} \exp\left(-\frac{1}{2|\Lambda|}\Sigma_{j,k}\Lambda^{jk}x_j x_k\right) \\
&= \frac{1}{(2\pi)^{n/2}\sigma_1\sigma_2\ldots\sigma_n\sqrt{|P|}} \\
&\quad \exp\left(-\frac{1}{2|P|}\Sigma_{j,k}P^{jk}\frac{x_j x_k}{\sigma_j \sigma_k}\right).
\end{aligned}
\tag{10.33}
$$

We see that if Λ is diagonal, the variables of the multidimensional normal distribution are independent, not just uncorrelated. The characteristic function of this distribution is

$$
\begin{aligned}
\phi(t_1, t_2, \ldots, t_n) &\equiv \mathbb{E}\left\{\exp\left(it^T x\right)\right\} \\
&= \exp\left(-\frac{1}{2}\Sigma_{j,k}\lambda_{jk}t_j t_k\right) = \exp\left(-\frac{1}{2}t^T \Lambda t\right).
\end{aligned}
\tag{10.34}
$$

The ellipsoid of concentration and the other ellipsoids obtained by replacing the right-hand side of Eq. 10.23 for the ellipsoid by any positive constant z are constant probability surfaces since $-\frac{1}{2}g(x_1, x_2, \ldots, x_n)$ corresponds to the exponent in the defining equation for f given above. The probability that a set of x_i lies outside this ellipse is given by the χ^2-distribution $\chi^2(z)$ for n degrees of freedom. Any conditional or marginal distribution of a multidimensional normal distribution is itself normal.

Let us look at transformations of variables. Suppose we have a density function of n variables, $f(x_1, x_2, \ldots, x_n)$, and wish to transform to new variables y_i. Then

$$f(x_1, x_2, \ldots, x_n)\, dx_1\, dx_2 \ldots dx_n = f(x_1, x_2, \ldots, x_n)|J|\, dy_1\, dy_2 \ldots dy_n. \quad (10.35)$$

Here $|J|$ is the magnitude (always taken greater than zero) of the Jacobian determinant:

$$J \equiv \frac{\partial(x_1, x_2, \ldots, x_n)}{\partial(y_1, y_2, \ldots, y_n)} \equiv \begin{pmatrix} \frac{\partial x_1}{\partial y_1} & \cdots & \frac{\partial x_n}{\partial y_1} \\ \vdots & & \vdots \\ \frac{\partial x_1}{\partial y_n} & \cdots & \frac{\partial x_n}{\partial y_n} \end{pmatrix} \quad (10.36)$$

If the transformation is a linear orthogonal transformation, then $|J| = 1$. In matrix notation, this occurs if $y = Cx$ and C is an orthogonal matrix ($CC^T = 1$, $T = $ transpose). The Jacobian is then the magnitude of the determinant of C^{-1}, which equals one.

Suppose we make a more general linear transformation from the variables x to the variables y by $y = Cx$, where we now allow the matrix C to be rectangular; i.e., allow there to be fewer y variables than x variables. If M is the moment matrix of the variable y, we can see easily that $M = C\Lambda C^T$. Consider the characteristic function $\phi(t) \equiv \mathbb{E}\{e^{it^T x}\}$. Let $t = C^T u$. Then $t^T x = u^T C x = u^T y$. Thus, the characteristic function, $\Psi(u)$, for the new distribution is given by

$$\begin{aligned} \Psi(u) &\equiv \mathbb{E}\{e^{iu^T y}\} = \mathbb{E}\{e^{it^T x}\} = \phi(t) \\ &= \exp\left(-\frac{1}{2}t^T \Lambda t\right) = \exp\left(-\frac{1}{2}u^T C\Lambda C^T u\right) \\ &= \exp\left(-\frac{1}{2}u^T M u\right). \end{aligned} \quad (10.37)$$

Thus, we see this is the characteristic function of a normal distribution with moment matrix M. Since we learned in Chap. 7 that the characteristic function uniquely determines the distribution, we see that we have shown the following theorem:

Any number of linear functions of normally distributed variables are themselves normally distributed.

In this chapter we have examined the general properties of two- and n-dimensional probability distributions. We have examined the properties of correlation coefficients. We have also examined the generalization of the normal distribution to two and many dimensions.

10.3   Exercises

10.1 Consider a two-dimensional density function proportional to

$$\exp\left(-(2x^2 + 2x + 1/2 + 2xy + 8y^2)\right).$$

(a) Find m_1, m_2, σ_1, σ_2, ρ, λ_{ij}.
(b) Write the correctly normalized density function.
(c) Make a linear non-singular transformation such that the new variables are uncorrelated. Write the density function in terms of these new variables.

10.2 Suppose we have a collection of diatomic molecules that we excite with laser beam pulses. The molecules can ionize giving up an electron or dissociate giving up an ion. At each pulse, we measure the number of ions (x) and the number of electrons (y) in appropriate detectors for each. We wish to ask whether there are some joint processes yielding both dissociation and ionization. Show that we can do this by measuring the fluctuations. Specifically, show that if we measure $\mathbb{E}\{x_i y_i\} - \mathbb{E}\{x_i\}\mathbb{E}\{y_j\}$, then we measure a quantity that is zero if there are no correlations and whose non-zero value is proportional to the correlation coefficient ρ. You may assume x and y are large enough at each pulse that we may use the two-dimensional normal distribution. This method is known as covariance mapping.

Chapter 11
The Central Limit Theorem

Abstract We have already seen that many distributions approach the normal distribution in some limit of large numbers. We will now discuss a very general theorem on this point, the central limit theorem. The normal distribution is the most important probability distribution precisely because of this theorem. We also will find that occasionally in regions of physical interest the assumptions fail and the distribution is not approached. We will also discuss a remarkable theorem on large numbers, Khintchine's law of the iterated logarithm.

11.1 The Lindeberg Criterion

Let $y = x_1 + x_2 + \cdots + x_n$, where the x_j are independent random variables. For any fixed n, of course, we have

$$\sigma_y^2 = \sigma_{x_1}^2 + \sigma_{x_2}^2 + \cdots + \sigma_{x_n}^2, \quad \overline{y} = \overline{x_1} + \overline{x_2} + \cdots + \overline{x_n}. \tag{11.1}$$

For a broad range of distributions, $F((y - \overline{y})/\sigma_y)$ approaches a normal distribution as n approaches infinity. This is the central limit theorem. In our discussion of the normal distribution in Chap. 6, we gave a simple proof of the central limit theorem for an important but restricted case. It would be useful for you to review that discussion to get a feeling for how this theorem comes about. It is, in fact, true for a wide variety of distributions and there are many criteria for the central limit theorem to be true. One of the more standard ones is the Lindeberg criterion. Let

$$\begin{aligned} \mu_k &= x_k \quad \text{if } |x_k - \overline{x_k}| \le \epsilon \sigma_y, \ k = 1, 2, \ldots, n \\ &= 0 \quad \text{if } |x_k - \overline{x_k}| > \epsilon \sigma_y. \end{aligned} \tag{11.2}$$

where ϵ is an arbitrary fixed number. The central limit theorem can be shown to be true if variance $(\mu_1 + \mu_2 + \cdots \mu_n)/\sigma_y^2 \to 1$ as $n \to \infty$. This criterion asks that

© Springer Nature Switzerland AG 2020

B. P. Roe, *Probability and Statistics in the Physical Sciences*,
Undergraduate Texts in Physics,
https://doi.org/10.1007/978-3-030-53694-7_11

Fig. 11.1 Distribution of the sum of five uniformly distributed random numbers compared to a normal distribution

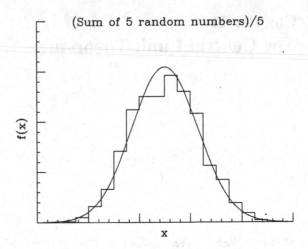

the fluctuation of no individual variable (or small group of variables) dominates the sum.

The theorem is also true if all the x_j are chosen from the same distribution and $\overline{x_j}$ and σ_j^2 exist.

What are some examples of this theorem? Suppose we make many trials from a Poisson distribution. The distribution of successes must approach the normal distribution. We have already seen this to be true.

Consider a variable chosen from a uniform distribution. That is, consider x_j to be a number chosen randomly between 0 and 1. In that case, the distribution $F((y - \overline{y})/\sigma_y)$, where $y = \sum_{j=1}^{n} x_j$, approaches a normal distribution. In fact, this is one way of generating a normal distribution on a computer using the usual uniform pseudorandom number generator present on most computers! See Figs. 11.1, 11.2, and 11.3. The horizontal scales are the same for each figure. Note that the width of the distribution narrows as n increases.

Next suppose we have a function g of n independent random variables x_i.

$$g(x_1, x_2, \ldots, x_n) = g(\overline{x_1}, \overline{x_2}, \ldots, \overline{x_n}) + \sum_{i=1}^{n} \frac{\partial g}{\partial x_i}\bigg|_{\overline{x}} (x - \overline{x_i}) + R. \qquad (11.3)$$

Often, if n is large, R is small and g is approximately a sum of n independent random variables. If this true and the central limit theorem applies, the distribution of g will approach a normal distribution. This is true in practice even if the variables have mild correlations. Since most measurements have many random causes of error, this is the justification for the usual assumption that measuring errors are normally distributed. We will see shortly some limitations on this.

We have seen that the sum of a large number of random variables often approaches a normal distribution. If a variable x is the product of a large number of factors, then,

Fig. 11.2 Distribution of the sum of 10 uniformly distributed random numbers compared to a normal distribution

Fig. 11.3 Distribution of the sum of 25 uniformly distributed random numbers compared to a normal distribution

subject to the tests above, it will follow a log-normal distribution. An example would be the output of a multistage photomultiplier tube. The output signal is the product of the input signal times the amplifications of each stage.

11.2 Failures of the Central Limit Theorem

Does the central limit theorem ever fail for real physical cases? The answer is yes! Consider the multiple scattering of a particle described previously. We had found

$$\sigma \equiv \sqrt{\overline{\theta^2}} \cong \frac{15.2}{p\beta} \sqrt{\frac{L}{L_R}}, \qquad \text{two-dimensional projection.} \qquad (11.4)$$

Fig. 11.4 Comparison of
angular distribution of
multiple scattering and that
of Rutherford scattering

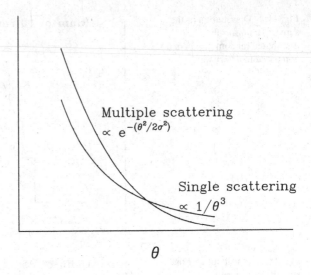

Multiple scattering proceeds largely through electromagnetic scattering off of the
nucleus for which the scattering formula for a single scattering is the Rutherford one:

$$f(\theta)\, d\theta\, dL = \frac{K\, d\theta\, dL}{\theta^3} \quad \text{for small angles.} \tag{11.5}$$

There is a clear problem here. The tails of the distribution fall off as $1/\theta^3$ for single
scattering. For multiple scattering,

$$f(\theta)\, d\theta\, dL \propto \exp\left(-\theta^2/2\sigma^2\right) d\theta\, dL. \tag{11.6}$$

If we go far enough out on the tail, the single scattering must dominate and the
distribution will not be normal. See Fig. 11.4.

Furthermore, for the Rutherford formula

$$\overline{\theta^2} = \int_0^\infty \theta^2 f(\theta)\, d\theta \sim log\, \theta|_0^\infty. \tag{11.7}$$

We might hope to avoid the difficulty in several ways. At large single scattering
angles, the Rutherford formula is, in fact, cut off by the nuclear form factor (the size
of the nuclei). This eventually cuts off the $1/\theta^3$ tail. However, this would imply that
the central limit theorem should only apply for scattering for which

$$\overline{\theta^2} >> \theta^2 \text{ cutoff.}$$

Most cases of interest do not satisfy this criterion.

In the above calculation of $\overline{\theta^2}$, we integrated to infinity, whereas the maximum angle is π. This turns out not to help for several reasons. In the first place, for many problems, π is effectively infinity. Second, if particles are scattered through large angles, we must consider the multiple valuedness of the angle and $\overline{\theta^2}$ becomes ill-defined.

Maybe the crossover is far out on the tail of the distribution and gets farther out as L increases. Let us see. Consider the crossover point, i.e., the point at which the distribution functions for Rutherford and multiple scattering are equal. Let

$$\theta \text{ crossover} \equiv \theta_c \equiv \sqrt{2} r \sigma , \quad r = \text{a number.} \tag{11.8}$$

Consider three-dimensional multiple scattering (it is easier to integrate) for fixed L. Let the distribution function for θ_c be $G(\theta_c)$.

$$G(\theta_c) = \int_{\theta_c}^{\infty} \left(\frac{1}{\sqrt{2\pi\sigma^2}} \right)^2 \exp\left(\frac{-\theta^2}{2\sigma^2} \right) d\theta_x \, d\theta_y , \quad \theta^2 = \theta_x^2 + \theta_y^2.$$

Change to polar coordinates $d\theta_x \, d\theta_y \rightarrow \theta \, d\theta \, d\phi$. The ϕ integration gives 2π.

$$G(\theta_c) = \frac{1}{2\pi\sigma^2} 2\pi \frac{1}{2} 2\sigma^2 e^{-r^2} = e^{-r^2}. \tag{11.9}$$

Consider the Rutherford scattering distribution function $F(\theta_c)$:

$$F(\theta_c) = \int_{\theta_c}^{\infty} \frac{KL}{\theta^3} d\theta = \frac{KL}{2\theta_c^2} = \frac{KL}{4r^2\sigma^2}. \tag{11.10}$$

But σ^2 is proportional to L and, therefore, we find $F(\theta_c)$ is independent of L. Hence, the crossover $F = G$ occurs at some value of r independent of L. The tail does not move farther out as L increases. For iron, $r = 2.027$ and $e^{-r^2} = 0.0164$ is the integral probability at the crossing point. Multiple scattering distributions in practice have a single scattering tail as shown in Fig. 11.5. The region near the crossover point is known as the plural scattering region. It is dominated by occurrences of a few (but not a large number) of collisions and is a difficult region to calculate.

Another example of the failure of the central limit theorem occurs in straggling, i.e., the spread in range of a stopping particle. When a charged particle such as a moderate energy muon travels in a medium, it undergoes a series of collisions with the atomic electrons, which slow it down. [We have chosen a muon because we wish to ignore strong interaction collisions. Electrons have additional energy loss mechanisms such as radiation and direct pair production that only become important at high energies ($\sim$100 GeV) for muons.] Furthermore, because of kinematics, the Rutherford scattering off the nucleus we considered for multiple scattering causes little energy loss, whereas the "knock-on" process of hitting atomic electrons causes energy loss but little angular change. You might try to work this out for yourself. The

Fig. 11.5 Comparison of angular distribution of multiple scattering and that of Rutherford scattering. The region of plural scattering is indicated

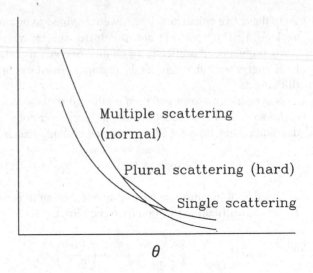

Multiple scattering (normal)

Plural scattering (hard)

Single scattering

θ

density function per collision for a particle with energy E to strike an electron and to give it an energy E' is

$$f(E,\ E')\,dE'dL = \frac{B}{(E')^2}\,dE'dL\,,\quad B = \text{a constant.} \tag{11.11}$$

The mean energy loss per collision is, therefore,

$$\overline{E'} = \int_{E'_{\min}}^{\sim E} \frac{B E'\,dE'}{E'^2}.$$

This is dominated by the many low-energy collisions since it has dE'/E' weighting in the integral. The variance involves

$$\overline{E'^2} = \int_{E'_{\min}}^{\sim E} (E')^2 B\,\frac{dE'}{E'^2}. \tag{11.12}$$

It is dominated by the few high-energy collisions. For 5-GeV muons, E' varies by nine orders of magnitude. Thus, events with $E' > E/10$ dominate the error, but correspond to very few collisions. Hence, the errors have the statistics of only a few events and are non-Gaussian. The distribution is furthermore skewed. To have a particle go further, it is necessary to have fewer collisions. However, one cannot have less than 0. These considerations are important in shielding calculations. The first person to calculate the probability of energy loss due to Coulomb interactions for a particle passing through some matter was Landau (1944). The distribution is known as the Landau distribution and the broad tail is known as the Landau tail.

Next consider the Cauchy (Breit–Wigner) distribution. We have found that the variance is infinite. Thus, the central limit theorem does not apply. We have seen that if x_i is a set of independent measurements and $x_{AV} = (1/n) \sum x_i$, then the distribution of x_{AV} is the *same* as that of x, i.e., the same Breit–Wigner with the same Γ. It does not get narrower. (This does not mean we cannot make a good measurement of E. In practice, we can measure the shape and thus measure the mass of a resonance.)

In Chap. 3 and at the beginning of this chapter, we considered the propagation of errors. The results there are all first-order results assuming the errors are small compared to the quantities. Often this is sufficient. Sometimes, however, non-linear effects are important. James (1983) looked at the data of an experiment that had apparently found a non-zero mass for the electron neutrino. They measured a quantity R, where

$$R = \frac{a}{\frac{d}{K^2 e}(b - c) - 2(1 - \frac{K^2 d}{Ke})a}. \tag{11.13}$$

We do not need to know the details here. It is sufficient that a, b, c, d, and e are measured, that K is fixed and that if $R < 0.420$, the neutrino must have non-zero mass. The experiment found that $R = 0.165$. The authors concluded that the error was $\sigma_R = 0.073$, using the linear propagation of errors described above. Since $R = 0.420$ is three standard deviations away, corresponding to a probability of one in a thousand for a normal distribution, it seemed that the neutrino must be massive.

However, the above formula is highly non-linear in the quantities actually measured and the linear approximation may be poor, especially since some of the errors were large. To test this James set up a Monte Carlo calculation assuming a, b, c, and d had independent normally distributed errors and looked at the distribution for R. He found that 1.5% of the time, the results gave $R > 0.42$, making the results far less certain than they seemed. James quoted 4%; the correction is quoted by Yost (1985).

I have observed that in many practical cases when one is measuring something, the measuring errors often have long tails for probabilities of less than 5 or 10%, owing either to effects similar to the ones considered above or owing to having a few large but low probability deviations.

11.3 Khintchine's Law of the Iterated Logarithm

We are now almost ready to turn to the inverse problem, trying to estimate a distribution from the data. Before doing so, let me close this chapter with a digression that I find amusing. This is a remarkable theorem on large numbers.

Consider a set of Bernoulli trials (recall the binomial distribution discussion). Let p be the probability of success; $q \equiv 1 - p$, the probability of failure; n, the number of trials; S_n, the number of successes. Let

$$S_n^* = (S_n - np)/(npq)^{\frac{1}{2}} = (S_n - m)/\sigma. \tag{11.14}$$

Fig. 11.6 Illustration of
Khintchine's law of the
iterated logarithm. S_n is the
number of successes in a
string of n Bernoulli trials
each having probability p of
success

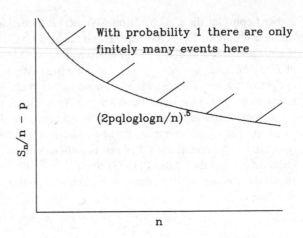

With probability 1 there are only finitely many events here

$S_n/n - p$

$(2pq\log\log n/n)^{.5}$

n

The law of the iterated logarithm (Khintchine) says that with probability 1

$$\limsup_{n\to\infty} \frac{S_n^*}{(2\log\log n)^{\frac{1}{2}}} = 1. \tag{11.15}$$

For $\lambda > 1$, as $n \to \infty$ there are only finitely times that

$$\left|\frac{S_n}{n} - p\right| > \lambda[2pq(\log\log n)/n]^{\frac{1}{2}}, \tag{11.16}$$

and, for $\lambda < 1$, there are infinitely many times (Fig. 11.6).

Although it is possible to have an infinite succession of all tails when flipping coins, it has probability 0. You should look back at our general discussion of probability in Chap. 1 with this theorem in mind.

In this chapter, we have examined the central limit theorem. Most distributions approach the normal distribution in some limit of large numbers. We have examined the conditions needed for this theorem to hold and have given physical examples of instances in which the theorem is valid and also examples of instances in which it is not valid. The central limit theorem is extremely important and useful, but it must never be used blindly.

11.4   Worked Problems

WP11.1 Show that multiple Coulomb scattering (m.s.) violates the Lindeberg criterion. Assume the single scattering is limited to angles between $\theta_{\min}$ and $\theta_{\max}$ and $\theta_{\max} > \sqrt{\overline{\theta^2}}$m.s. $= \sigma$.

Answer:
Recall that for single Coulomb scattering $f(\theta)d\theta = KLd\theta/\theta^3$. Therefore,

$$\text{variance}(\theta_i) = \int_{\theta_{\min}}^{\theta_{\max}} \theta^2 \frac{KLd\theta}{\theta^3} = A \log \frac{\theta_{\max}}{\theta_{\min}}.$$

$$\text{variance}\left(\sum \theta_i\right) = nA \log \frac{\theta_{\max}}{\theta_{\min}}.$$

Choose ψ such that $\psi = \epsilon\sigma$, where σ is the root mean square (r.m.s.) multiple scattering angle $= \sqrt{n}\sigma_i$. Let

$$u_i = \begin{cases} \theta_i, & \theta_i < \psi, \\ 0, & \theta_i > \psi. \end{cases} \tag{11.17}$$

$$\text{variance } u_i = A \log \frac{(\theta_{\max})_u}{\theta_{\min}} = A \log \frac{\psi}{\theta_{\min}}. \tag{11.18}$$

$$\frac{\text{variance} \sum u}{\text{variance} \sum x} = \frac{nA \log (\psi/\theta_{\min})}{nA \log (\theta_{\max}/\theta_{\min})}. \tag{11.19}$$

This does not approach one as n gets large and ϵ remains fixed (and <1). $\psi = \epsilon\sigma < \theta_{\max}$ as $\sigma < \theta_{\max}$ by assumption.

WP11.2 Some years ago, it was suggested that the fine structure constant α had a value $\alpha^{-1} = 2^{19/4}3^{-7/4}5^{1/4}\pi^{11/4} = 137.036082$. Experimentally, α^{-1} was measured to be 137.03611 ± 0.00021 at that time.
To see if this might be a numerical coincidence, Roskies (1971) examined by computer all numbers formed by a product $r = 2^a 3^b 5^c \pi^d$, where a, b, c, d are rational numbers of magnitude less than or equal to five with denominators less than or equal to 4. He found five other expressions giving α^{-1} within one standard deviation. How many would you expect statistically to be in this interval?
Note that $\log r = a \log 2 + b \log 3 + c \log 5 + d \log \pi$. You may assume each term is approximately uniformly distributed between its minimum and maximum value. Thus, the density function for $a \log 2$ is a constant between $-5 \log 2$ and $+5 \log 2$ and is 0 outside these limits. Using the central limit theorem, assume the sum of these four terms is approximately Gaussian with variance equal the sum of the variances of the individual terms. (Note this means $\log r$ is approximately Gaussian!)

Answer:
For a uniform distribution from 0 to ℓ (recall Worked Problem 1.1), $\sigma = \ell/\sqrt{12}$. For example, consider $a \log 2$. This can be considered approximately a uniform distribution from $-5 \log 2$ to $+5 \log 2$ and it, therefore, has $\sigma = 10 \log 2/\sqrt{12}$.

$$\log r = a \log 2 + b \log 3 + c \log 5 + d \log \pi.$$

$$\sigma_{\log r}^2 = \frac{100}{12}[(\log 2)^2 + (\log 3)^2 + (\log 5)^2 + (\log \pi)^2], \qquad (11.20)$$

$$= 46.56 \text{ or } \sigma = 6.82$$

$$\overline{\log r} = 0.$$

How many numbers are there for each of a, b, c, d?

$$\frac{1}{4} \; 30 \quad (\text{excluding whole numbers})$$

$$\frac{1}{3} \; 20, \qquad\qquad\qquad\qquad\qquad (11.21)$$

$$1 \; \underline{11},$$

$$61$$

$(61)^4 = 13.846 \times 10^6$ numbers.

$$\log 137 = 4.92,$$

$$\frac{\log 137}{\sigma} = \frac{4.92}{\sqrt{46.56}} = 0.721.$$

The quoted error at that time in the measured value of α was 0.00021.

$$d(\log r) \sim \frac{dr}{r} \sim \frac{2 \times 0.00021}{137} = 3.066 \times 10^{-6},$$

$$\frac{d(\log r)}{\sigma} = dx,$$

$$13.84 * 10^6 * f(x)\, dx = f_{\text{normal}}(0.721) * \frac{3 \times 10^{-6}}{6.82} \times 13.846 * 10^6$$

$$= 0.30 \times \frac{3 \times 10^{-6}}{6.82} \times 13.846 \times 10^6 = 1.87 = \text{expected number}.$$

The probability of obtaining 6 if we expect 1.87 is given by the Poisson distribution

$$P_n = \frac{e^{-\lambda}\lambda^n}{n!} = 0.009 \sim 1\%. \qquad (11.22)$$

11.5   Exercises

11.1 Show that the Lindeberg criterion is satisfied for the binomial distribution and, therefore, the binomial distribution does approach the normal distribution for large n. (Let each x_i be a single trial.)

References

James F (1983) Probability, statistics and associated computing techniques. In: Ferbel T (ed) Techniques and concepts of high energy physics II. Plenum Press, New York

Landau L (1944) On the energy loss of fast particles by ionization. J Phys USSR 8:201

Roskies R (1971) Letters to the editor. Phys Today 9

Yost G (1985) Lectures on probability and statistics. Technical report LBL-16993, Lawrence Berkeley Laboratory Report

Chapter 12
Choosing Hypotheses and Estimating Parameters from Experimental Data

Abstract First, we will discuss the important lemma of Fisher, which is the basis for our use of degrees of freedom when estimating parameters by making fits to experimental data. We will then discuss the maximum likelihood method and quote some general theorems concerning the efficiency and bias of estimates. Finally, we will examine sufficient statistics, and the testing and comparing of hypotheses using likelihood ratios.

12.1 Lemma of Fisher

Let x_1, x_2, $\ldots$, x_n be normally distributed, independent variables of mean zero and variance σ^2.

1. Consider a linear orthogonal transform from x_1, x_2, $\ldots$, x_n to y_1, y_2, $\ldots$, y_n, i.e., $y_i = \Sigma_j c_{ij} x_j$ or, in matrix notation, $y = Cx$, where C is an orthogonal matrix ($C^T = C^{-1}$). Thus, $(C^{-1})_{lm} = c_{ml}$. We have seen in the previous section that the joint distribution of y_1, y_2, $\ldots$, y_n is normal. Furthermore,

$$\mathbb{E}\{y_i y_k\} = \sigma^2 \sum_{j=1}^{n} c_{ij} c_{kj} = \sigma^2 \sum_{j=1}^{n} c_{ij} (c^{-1})_{jk} = \sigma^2 \delta_{ik}, \tag{12.1}$$

where δ is the Kronecker δ, 1 if the two indices are the same and 0 otherwise. Therefore, the y_i are independent. The normalization is also preserved. Then

$$R^2 = \sum_{i=1}^{n} y_i^2 = \sum_{i=1}^{n} \left(\sum_{j=1}^{n} c_{ij} x_j \right)^2 = \sum_{i=1}^{n} \sum_{j=1}^{n} \sum_{k=1}^{n} c_{ij} c_{ik} x_j x_k, \tag{12.2}$$

B. P. Roe, *Probability and Statistics in the Physical Sciences*,
Undergraduate Texts in Physics,
https://doi.org/10.1007/978-3-030-53694-7_12

$$R^2 = \sum_{j=1}^{n} \sum_{k=1}^{n} \delta_{jk} x_j x_k = \sum_{i=1}^{n} y_i^2 = \sum_{i=1}^{n} x_i^2. \tag{12.3}$$

Thus, this transformation acts like an orthogonal rotation of an orthogonal coordinate system to another orthogonal coordinate system.

2. Suppose we have only part of an orthogonal transformation given. That is, imagine only y_1, y_2, ..., y_p are given in the above transformation, where $p < n$ and for these p variables, $\Sigma_{i=1}^{n} c_{ij} c_{kj} = \delta_{ik}$. Then by the standard Gram–Schmidt process, we can always add $n - p$ more y variables to make a complete orthogonal transform. In this process, consider a new vector $\mathbf{V_0}$. Let

$$\mathbf{V} = \mathbf{V_0} - \sum_{i=1}^{p} \frac{\mathbf{V} \cdot \mathbf{y_i}}{\mathbf{y_i} \cdot \mathbf{y_i}} \times \mathbf{y_i}$$

If $\mathbf{V}$ is not identically zero, it can be normalized and added to the previous p variables and renormalized as desired. Continuing in the same way $n - p$ variables can be added.

The lemma of Fisher then states:

$$S^2 \equiv \sum_{i=1}^{n} x_i^2 - \sum_{j=1}^{p} y_j^2 = \sum_{j=p+1}^{n} y_j^2. \tag{12.4}$$

is the sum of $n - p$ independent normal variables with mean zero and variance σ^2. S^2/σ^2 is, thus, distributed in accordance with the chisquare distribution in $n - p$ degrees of freedom and is independent of y_1, y_2, ..., y_p.

If the mean is not zero, then the above lemma still holds if we replace x_i by $x_i - m_{x_i}$ and y_j by $y_j - m_{y_j}$. Thus, if x_1, x_2, ..., x_n are independent normally distributed variables with mean m_{x_i} and variance σ^2, then a linear orthogonal transform produces a set of variables y_1, y_2, ..., y_n that are independent, normally distributed with variance σ^2, and the orthogonal transform applied to the means of the x_i's gives the means of the y_j's.

What is the significance of this lemma? Suppose we have a problem in which we have a histogram with $n = 10$ bins and we wish to fit a curve with three parameters to it. If the conditions are linear, then these three parameters correspond to three linear equations or $p = 3$. The parameters can be considered as the first three of a transformed set of variables. We obtain a chi-square distribution with seven degrees of freedom. With no parameters, we had a chi-square distribution with ten degrees of freedom. We have lost one degree of freedom for each parameter that we fit.

We will explore these concepts further in Chaps. 13–17, but this lemma is the basic justification for the behavior of degrees of freedom when fitting parameters.

As a comment, let me note that sometimes you may have a matrix in which the set of eigenvalues is known and you wish to find the eigenvectors. There is a newly discovered method of doing this which simplifies the task (Denton et al. 2019).

12.2   Maximum Likelihood Method

This is a very powerful method for estimating parameters. Suppose we have a density function, which is a function of the vector $\mathbf{x}$, $f(\mathbf{x}|\alpha_1, \alpha_2, \ldots, \alpha_s)$, where $\mathbf{x}$ is the vector of measured variables and $\alpha_1, \alpha_2, \ldots, \alpha_s$ are s parameters. Suppose we now perform n independent experiments, obtaining $\mathbf{x}_1, \mathbf{x}_2, \ldots, \mathbf{x}_n$. Then

$$\mathcal{L} = f(\mathbf{x}_1) f(\mathbf{x}_2) \ldots f(\mathbf{x}_n) \tag{12.5}$$

is the density function for obtaining this set of events if $\boldsymbol{\alpha}$ is fixed. We call $\mathcal{L}$ the likelihood of the experimental result.

The maximum likelihood method consists in finding an estimate of the parameters, $\boldsymbol{\alpha}^*$, which maximizes $\mathcal{L}$. Since the maximum of $\mathcal{L}$ is also the maximum of $\log \mathcal{L}$, we usually maximize the latter function. Our set of likelihood equations is then

$$w = \log \mathcal{L} = \sum_{i=1}^{n} \log f(\mathbf{x}_i|\alpha_1, \alpha_2, \ldots, \alpha_s),$$

$$\frac{\partial w}{\partial \alpha_j} = \sum_{i=1}^{n} \frac{1}{f(\mathbf{x}_i)} \frac{\partial f(\mathbf{x}_i)}{\partial \alpha_j} = 0. \tag{12.6}$$

Before quoting the main maximum likelihood theorem, we must define some jargon.

We wish to make the best estimate of $\boldsymbol{\alpha}$, which we define to be the estimate with least variance. (In future chapters, we will consider a few cases in which it is desirable not to choose the least variance solution.) Consider a single parameter α and assume that each experiment has a single x. Transform from

$$f(x_1|\alpha) f x_2|\alpha) \ldots f(x_n|\alpha) \, dx_1 \ldots dx_n$$

to

$$g(\alpha^*|\alpha) h(\xi_1, \ldots, \xi_{n-1}, \alpha^*|\alpha) \, d\alpha^* \, d\xi_1 \ldots d\xi_{n-1}.$$

Suppose that for almost all values of x, α^*, $\xi_1, \ldots, \xi_{n-1}$, the partial derivatives $\partial f/\partial \alpha$, $\partial g/\partial \alpha$, and $\partial h/\partial \alpha$ exist for every α in A and that

$$\left|\frac{\partial f}{\partial \alpha}\right| < F_0(x), \left|\frac{\partial g}{\partial \alpha}\right| < G_0(\alpha^*), \left|\frac{\partial h}{\partial \alpha}\right| < H_0(\xi_1 \ldots \xi_{n-1}, \alpha^*), \tag{12.7}$$

where $F_0, G_0, \alpha^* G_0$, and H_0 are integrable over the whole space of variables x and α^*, $\xi_1 \ldots \xi_{n-1}$, respectively. Then this is regular estimation of the continuous type and α^* is a regular estimate of α.

Suppose for this type of estimation we use some criteria, not necessarily maximum likelihood, to obtain an estimate, α^*, whose expectation value is given by

$$\mathbb{E}\{\alpha^*\} = \alpha + b(\alpha),$$ (12.8)

where b is the bias of α^*. Then it can be shown that

$$\mathbb{E}\left\{(\alpha^* - \alpha)^2\right\} \geq \frac{(1 + db/d\alpha)^2}{n \int_{-\infty}^{\infty} (\partial \log f/\partial \alpha)^2 \, f(x\,|\,\alpha) \, dx}.$$ (12.9)

An estimator is said to be *consistent* if the estimate α^* converges in probability to the true value α as the amount of data increases. If the estimate α^* has $\mathbb{E}\{\alpha^*\} = \alpha$, the estimate is called *unbiased*. The bias depends on the chosen metric, i.e., if α^* is an unbiased estimator of α, then $(\alpha^*)^2$ is not in general an unbiased estimator for α^2. For unbiased estimates, we have

$$\mathbb{E}\left\{(\alpha^* - \alpha)^2\right\} \geq \left[n \int_{-\infty}^{\infty} \left(\frac{\partial \log f}{\partial \alpha}\right)^2 f(x\,|\,\alpha)\,dx\right]^{-1}.$$ (12.10)

We can use this minimum value on the right-hand side of Eq. 12.10 as a standard to compare with all other estimates. The *efficiency* of an estimate (e) is the ratio of this limiting variance to that of the estimate in question. If $e \to 1$ as the number of observations $n \to \infty$, the estimate is *asymptotically efficient*. Suppose we have a density function $f(x)$ and we make n observations. We are trying to estimate a single parameter α. Make a transformation of variables $x_1, \ldots, x_n \to \zeta_1, \zeta_2, \ldots, \zeta_{n-1}, \alpha^*$, where α^* is our estimate for α. Suppose we can find a transformation such that

$$f(x_1)f(x_2)\ldots f(x_n) = g(\alpha^*\,|\,\alpha)h(\zeta_1, \zeta_2, \ldots, \zeta_{n-1}, \alpha^*),$$ (12.11)

i.e., such that h does not depend on the real value of the parameter α. In this case, α^* summarizes all of the information in the sample relevant to α. α^* is then called a *sufficient statistic*. When possible, it is desirable to base our estimates on sufficient statistics. We can now proceed to our theorem.

The following theorem justifies the likelihood method for the case of a single measured vector **x**, (one dimensional) and a single parameter α (one dimensional). Suppose the following conditions are satisfied:

(a) For almost all x, the derivatives

$$\frac{\partial \log f}{\partial \alpha}, \quad \frac{\partial^2 \log f}{\partial \alpha^2}, \quad \frac{\partial^3 \log f}{\partial \alpha^3}$$ (12.12)

exist for all α in a non-degenerate interval A.

(b) For every α in A,

$$\left|\frac{\partial f}{\partial \alpha}\right| < F_1(x), \quad \left|\frac{\partial^2 f}{\partial \alpha^2}\right| < F_2(x), \quad \left|\frac{\partial^3 \log f}{\partial \alpha^3}\right| < H(x),$$ (12.13)

where F_1 and F_2 are integrable in the interval $(-\infty, \infty)$ and

$$\int_{-\infty}^{\infty} H(x) f(x)\, dx < M, \tag{12.14}$$

where M is independent of α.

(c) For every α in A,

$$\int_{-\infty}^{\infty} \left(\frac{\partial \log f}{\partial \alpha}\right)^2 f\, dx \tag{12.15}$$

is finite and greater than zero.

Then the likelihood equation has a solution α^*, which converges in probability to the true value, α_0, as $n \to \infty$. Assume now that we do a set of estimates. Each estimate involves n measurements. Then the distribution of α^* over the different estimates is asymptotically normal and α^* is asymptotically efficient as $n \to \infty$. Thus, let

$$k^2 = \mathbb{E}\left\{\left(\frac{\partial \log f}{\partial \alpha}\right)^2_{\alpha=\alpha_0}\right\},$$

$$y = k\sqrt{n}(\alpha^* - \alpha_0). \tag{12.16}$$

y is asymptotically normal $(0, 1)$. The variance of α^* is asymptotically

$$\left(n\mathbb{E}\left\{\left(\frac{\partial \log f}{\partial \alpha}\right)^2\right\}\right)^{-1}. \tag{12.17}$$

This theorem can be extended to problems involving several parameters and to problems involving correlated events, i.e., $f(x_1, \ldots, x_n) \neq f(x_1) f(x_2) \ldots f(x_n)$.

In addition to the above main theorem, the following three results can also be proven (Cramer 1946).

1. If an efficient estimate α^* exists, the likelihood equation will have a unique solution, α^*.
2. If a sufficient statistic exists, any solution of the likelihood equation will be a function of the sufficient statistic. Thus, whenever possible, the estimate will be based on a sufficient statistic.
3. Consider a grouped continuous or a discrete distribution so that the results are divided into r bins. The maximum likelihood method (if the conditions of the main theorem are met), or, in fact, any asymptotically normal and asymptotically efficient estimate of the s parameters, will have χ^2 asymptotically distributed according to a chi-square distribution in $r - s - 1$ degrees of freedom.

There are very useful other forms for the variance. In Chap. 15, we will show that

$$\sigma^2 = \left(n\mathbb{E}\left\{ \left(\frac{\partial \log \dot{f}}{\partial \alpha} \right)^2 \right\} \right)^{-1} = \left(\mathbb{E}\left\{ \left(\frac{\partial w}{\partial \alpha} \right)^2 \right\} \right)^{-1}$$

$$= \left(\mathbb{E}\left\{ \frac{-\partial^2 w}{\partial \alpha^2} \right\} \right)^{-1}, \quad w = \log \mathcal{L}. \tag{12.18}$$

Suppose we make a series of tests yielding N real random variables $\mathbf{X}$ and wish to estimate a parameter θ. Call Ω_θ the set of all allowed values of $\mathbf{X}$, which might depend on the value of θ.

The Darmois Theorem finds strong restrictions on probability density functions, which admit sufficient statistics independent of the number N of observations (James 2006). For a single variable θ, the theorem finds that if the data set $\mathbf{X}$ admits a sufficient statistic, then the probability function is an exponential:

$$f(\mathbf{X}|\theta) = e^{[\beta(\alpha)a(\theta)+\gamma(\mathbf{X})+c(\theta)]}, \tag{12.19}$$

and inversely, $\mathbf{X}$ admits a sufficient statistic for all $N > 1$ but only if Ω_θ is independent of θ, and if

$$\mathbf{X} \Rightarrow (R, X_2, \ldots, X_N), \quad \text{where } R = \sum_{i=1}^{N} \beta(X_i), \tag{12.20}$$

where R is one-to-one continuously differentiable for all $\mathbf{X}$. R is sufficient for θ. The multidimensional exponential form is

$$f(\mathbf{X}|\theta) = e^{\sum_{j=1}^{S} \beta_j(\mathbf{X})a_j(\theta)+\gamma(\mathbf{X})+c(\theta)} \text{ and } R_j = \sum_{i=1}^{N} \beta_j(X_i), j = 1, \ldots, S, \tag{12.21}$$

is one possible set of S jointly sufficient statistics for $\boldsymbol{\theta}$.

It is important to use properly normalized probabilities when doing a maximum likelihood fit. One might think that the normalization is not important, since it is an additive constant after taking the logarithm. However, it can be very important if the normalization depends on the parameter α that is being estimated. For example, suppose we are measuring radiation as a function of frequency for a system and trying to measure the magnetic moment and other parameters of the system by looking at the shape of the frequency spectrum of the radiation. If we just use the differential cross section for the probability without normalization, we will get an incorrect answer. The reason is that the cross section will increase as the magnitude of the magnetic moment increases and every one of our probabilities will increase. Hence, the higher the magnetic moment hypothesized, the higher the likelihood! To look at the shape dependence, it is important to normalize so that the integral of the probability distribution does not depend on the parameter of interest.

The maximum likelihood method is often used for fitting data in the form of histograms. However, it is also used for more general curve fitting. For this latter use, problems can occur if the measurement errors are not normal. We will discuss these problems in Sect. 15.2 of Chap. 15 and, in Chaps. 17 and 19, we will describe some estimation methods that are more suitable for this kind of problem.

It is worth noting that although estimating the median for a distribution can often be more difficult computationally than the mean, the resulting estimator is generally more robust, as it is insensitive to the exact shape of the tails of a distribution.

12.3 Testing and Comparing Hypotheses

In this section, we will consider several problems. Suppose we are trying to test a hypothesis H_0, traditionally called the null hypothesis and have made a series of n tests which yielded results $x_1, x_2, \ldots, x_n = \mathbf{x}$. The first problem assumes that H_0 is fixed (simple hypothesis) with no parameters to set. The second problem (composite hypothesis) has parameters to be set. The third problem involves a comparison with one or more alternative hypotheses. A test statistic $\mathbf{t}(\mathbf{x})$ is used to examine the hypothesis. Each hypothesis i has its own probability density function, $g(\mathbf{x}|H_i)$.

We define a critical region, such that if $g(\mathbf{x}|H_0)$ falls in this region H_0 is rejected. The significance level of the test s is the probability that $\mathbf{t}$ falls in this region assuming H_0 is correct. The complement of the critical region is called the acceptance region (Fig. 12.1). Two kinds of error are distinguished. An error of the first kind (type 1 error) occurs if H_0 is true, but is rejected. An error of the second kind (type 2 error) occurs if H_0 is false, but is accepted.

The power of the test is one minus the fraction of errors of the second kind. For a given experimental result for H_0, the p-value is defined as the probability of getting a result equally or less probable than the result obtained. The p-value is a function of the data, and is, therefore, itself a random variable. If the hypothesis used to compute the p-value is true, then for continuous data p will be uniformly distributed between zero and one. Note that the p-value is not the probability for the hypothesis; in frequentist statistics, this is not defined. When searching for a new phenomenon, one tries to reject the hypothesis H_0 that the data are consistent with known processes. If the p-value of H_0 is very low, then we reject the hypothesis. To make it more familiar to many scientists it is convenient to distort the flat distribution into a normal distribution. This is easily done by setting the inverse of the normal distribution function $D^{-1} = 1 - p$. In particle physics 5σ is taken as the level needed to qualify a new result as a discovery. This corresponds to a p-level of 2.87×10^{-7}. (There are several reasons such a rigorous criterion is selected. The look elsewhere effect, which we will discuss later, occurs when the result could have occurred in any of a number of places and must be derated by that number. Also, the errors are often not Gaussian, but have long tails (kurtosis) or we sometimes have ill-defined systematic errors.)

Suppose there are alternative hypotheses and we are examining events, some of which correspond to the null hypothesis and some to alternative hypotheses. The efficiency of a selection ϵ is the fraction of null hypothesis events accepted. The

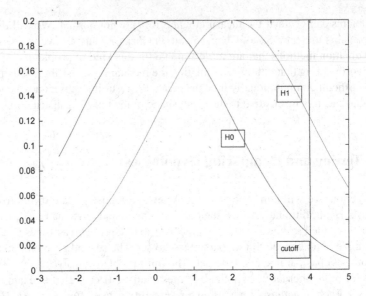

Fig. 12.1 The plots show the pdf's of H0 and H1 versus the test statistic $\mathbf{t(x)}$, the horizontal axis. The critical region is chosen as $\mathbf{t(x)} > 4$, that is, if $\mathbf{t(x)} > 4$, for H_0 or H_1, that hypothesis is rejected. The probability of that happening is the integral of the appropriate pdf's, $g(\mathbf{t(x)} > 4|H_i)$ with $i = 0$ or 1. If the experimental result $\mathbf{t(x_{exp})} < 4$, for both $H0$ and $H1$, the likelihood ratio is the ratio of the probabilities, $\Lambda = \mathcal{L}[g(\mathbf{t(x_{exp})}|H0)]/\mathcal{L}[g(\mathbf{t(x_{exp})}|H1)]$

purity of a selection π is the number of null hypothesis events accepted divided by the total number of events selected.

An extension of the result in the last section is that given essentially modest integrability and continuity conditions, if H_0 is true and there is one alternate H_1, the acceptance region giving the highest power for a given selection efficiency is determined by the likelihood ratio $g(\mathbf{x}|H_0)/g(\mathbf{x}|H_1) > c$, where c is chosen to get the desired efficiency (Brandt 1992, 1997). This can be thought of as setting the test statistic t to this ratio. To use this result, it is necessary to know $g(\mathbf{x}|H_0)$ and $g(\mathbf{x}|H_1)$ which is often done by Monte Carlo methods. Some linear approximations are discussed in Cowan (1998).

It can be shown that for a simple or composite hypothesis, the log of the likelihood function approaches a normal distribution for large n. This is also true when comparing two simple hypotheses.

Suppose we have a maximum likelihood fit with s parameters and want to test whether certain restrictions between these parameters are obeyed. Let

$$\Lambda = \frac{\mathcal{L} \text{ max without restrictions}}{\mathcal{L} \text{ max with restrictions}}. \tag{12.22}$$

If the restrictions above are obeyed, it can be shown (Lyons 2017; Wilks 1946) that if $\mathcal{L}$ satisfies mild continuity restrictions (similar to the ones in the main theorem quoted earlier in this section), then as n approaches infinity, the distribution of $2 \log \Lambda$ approaches the chi-square distribution with $s - r$ degrees of freedom, where s is the total number of parameters and r is the effective number of parameters after the restrictions are made. If a sufficient statistic exists, the likelihood ratio test will be based on the sufficient statistic.

The likelihood ratio is sometimes called the betting odds for A against B. As we will see later, this can be a very useful concept. However, it is sometimes misleading. Suppose, for a specific example, A has a probability of 20% and B of 80%. We should be leery of betting 4 to 1 on B since B, if true, is an exceptionally lucky fit and A is quite an acceptable fit. On the other hand, if A has a probability of 10^{-6} and B has a probability of 10^{-8}, we would hesitate to bet 100 to 1 on B. Here neither A nor B are likely and the ratio is not relevant. We see that, although it can be useful, this one number cannot completely summarize two results.

References

Brandt S (1992) Date, analyse, 3rd edn. BI-Wissenschaftsverlag, Mannheim
Brandt S (1997) Statistical and computational methods in data analysis. Springer, New York
Cowan G (1998) Statistical data analysis. Clarendon Press, Oxford
Cramer H (1946) Mathematical methods of statistics. Princeton University Press, Princeton
Denton P, Parke S, Tao T, Zhang X (2019) Eigenvectors from eigenvalues. arXiv.1908.03795
James F (2006) Statistical methods in experimental physics, 2nd edn. World Scientific Publishing Co. Pte. Ltd., Singapore
Lyons L (2017) A paradox about the distribution of likelihood ratios? arXiv:1711.00775
Wilks S (1946) Mathematical statistics. Princeton University Press, Princeton

Chapter 13
Method of Least Squares (Regression Analysis)

Abstract We are now ready to look at particular methods of estimating parameters from data. As we will see, there is a very sophisticated methodology that has developed for these problems and, in many cases, the estimation process can be automated and turned into almost a crank-turning operation. Nonetheless, it is very important to understand in detail what we are doing, and we will discuss problems that have arisen when people were not sufficiently careful. We will discuss how the error in the estimation of a parameter using a χ^2 fit can be a normal distribution, while the χ^2 as a whole has a χ^2 distribution. We discuss some problems for fitting histograms and taking the theory at the center point, and an approximate correction called Shepard's corrections for grouping. Correlation coefficients are often poorly determined in fits and we discuss a suggested modification. Finally, we discuss a method for putting together several uncorrelated probabilities.

13.1 Least Squares Analysis

Suppose we have a histogram divided into r bins. Let bin i have v_i events with a total number of n events in the sample.

$$\sum_{i=1}^{r} v_i = n. \tag{13.1}$$

An initial consideration is how to choose the bins. There is no need for the bins to be of equal width and an optimal method for dividing into bins is to choose the bins to have equal probability based on an initial estimate of the parameters. Furthermore, if possible, choosing the size to have ten or more events makes a normal distribution approximation reasonable. With few events, the distribution is Poisson.

For the general case, imagine the distribution is known to be such that an individual event has a probability p_i of falling into the ith bin.

$$\sum_{i=1}^{r} p_i = 1. \tag{13.2}$$

© Springer Nature Switzerland AG 2020
B. P. Roe, *Probability and Statistics in the Physical Sciences*,
Undergraduate Texts in Physics,
https://doi.org/10.1007/978-3-030-53694-7_13

Furthermore, we suppose $p_i \ll 1$ for all i and imagine that p_i is a function of s parameters, α_j, whose value we wish to estimate $(s < r)$. We define

$$\chi^2 = \sum_{i=1}^{r} \left(\frac{(v_i - np_i)^2}{np_i} \right). \tag{13.3}$$

If p_i is not small, then we change the denominator to $np_i q_i$, where $q_i = 1 - p_i$.

If there were no parameters to estimate, it can be shown that as n approaches infinity, χ^2 is distributed in a χ^2 distribution with $r - 1$ degrees of freedom. We have lost one degree of freedom because of the constraint that the total number of events is fixed.

For the present case, there are s parameters. The method of least squares consists of choosing the estimates, α_j^*, of the α_j in such a manner that χ^2 is minimized. Thus, we might set

$$-\frac{1}{2} \frac{\partial \chi^2}{\partial \alpha_j} = \sum_{i=1}^{r} \left(\frac{v_i - np_i}{p_i} + \frac{(v_i - np_i)^2}{2np_i^2} \right) \frac{\partial p_i}{\partial \alpha_j} = 0 \quad \text{for } j = 1, 2, \dots, s. \tag{13.4}$$

However, χ^2 is not a probability. The log of the likelihood indeed has $e^{-\chi^2/2}$, but there are also normalization terms. If, for example, some of the parameters affect the number of events expected in a histogram there is problem which periodically causes confusion, even in published papers.

13.2 Problem with the Use of χ^2 if the Number of Events Is Not Fixed

Consider a simple toy model. Suppose we have a data set with only two bins. Theory predicts that the expected value of N, the number of events in the bin, should be the same for each bin, and that the bins are uncorrelated. Let x_1 and x_2 be the number of events experimentally found in the two bins. The variance (σ^2) is N for each bin, $(\sigma = \sqrt{N})$.

$$\chi^2 = \frac{(N - x_1)^2}{\sigma^2} + \frac{(N - x_2)^2}{\sigma^2}. \tag{13.5}$$

We want to find the minimum, $\frac{\partial \chi^2}{\partial N} = 0$. Call term 1, the derivative with respect to the numerators of the χ^2.

$$\text{Term 1} = 2\frac{(N - x_1 + N - x_2)}{N} = 2\left(1 - \frac{x_1}{N}\right) + 2\left(1 - \frac{x_2}{N}\right). \tag{13.6}$$

If we ignore the derivative of the denominator, then Term 1 = 0, is solved by

$$N = (x_1 + x_2)/2. \tag{13.7}$$

Call this the naive solution.

Call Term 2 the derivative with respect to the denominator of the χ^2

$$\text{Term } 2 = -\frac{(N - x_1)^2 + (N - x_2)^2}{N^2}. \tag{13.8}$$

Term 2 is negative and O(1). The only way that Term 1 + Term 2 = 0 is for Term 1 to be positive. This means that the χ^2 solution must have N greater than the naive value. Although Term 1 is O(1), x_1/N and x_2/N are O(1/N). N is pulled up as the fit wants to make the fractional errors larger.

Suppose, incorrectly, one had used as a variance the number of data events rather than the expected number of events (N), which some have tried, then

$$\chi^2 = \frac{(N - x_1)^2}{x_1} + \frac{(N - x_2)^2}{x_2} \tag{13.9}$$

Upon taking derivatives then instead of the desired $N = 0.5(x_1 + x_2)$, one gets the averages of the inverses which is lower than the naive result:

$$\frac{2}{N} = \frac{1}{x_1} + \frac{1}{x_2} \tag{13.10}$$

This result is actually not as bad as it may seem at first. Define ϵ by $x_2 = x_1(1 + \epsilon)$. Then the solution (13.10) differs from (13.7) only by a term of order ϵ^2. For large x's, the solutions are quite close. A large number of analyses in current experiments use the number of data events as the variance and do a χ^2 fit. The residual small bias, might be explored by using Monte-Carlo events.

However, if one correctly uses as variance the number of events expected, the problem remains. The problem occurs because the χ^2 is not a probability. The likelihood ($\mathcal{L}$) is the probability density function for the two bins assuming each bin has a normal distribution. (This requires N is not too small). The log of the likelihood has normalization factors in addition to $-\chi^2/2$. It is these missing normalization factors that cause the problem. We noted in Sect. 12.1 the importance for likelihood itself to have proper normalization if N was not fixed.

$$\mathcal{L} = \frac{1}{\sqrt{2\pi\sigma^2}} \frac{1}{\sqrt{2\pi\sigma^2}} e^{-(N-x_1)^2/(2\sigma^2)} e^{-(N-x_2)^2/(2\sigma^2)}. \tag{13.11}$$

For $\sigma^2 = N$, the log of the likelihood is

$$\ln \mathcal{L} = -\ln(2\pi) - \ln N - \chi^2/2. \tag{13.12}$$

Let Term 3 be the derivative of the normalization.

$$\text{Term } 3 = -\frac{1}{N}.\tag{13.13}$$

The derivative of the $\ln \mathcal{L}$ is Term 3 $-$ (Term 1)/2 $-$ (Term 2)/2.

$$\text{Term } 3 - \text{Term } 2/2 = -\frac{1}{N} + \frac{(N - x_1)^2 + (N - x_2)^2}{2N^2} = \frac{-2N + (N - x_1)^2 + (N - x_2)^2}{2N^2}$$
$$\tag{13.14}$$

Since the expectation value $\mathbb{E}(N - x_1)^2 + \mathbb{E}(N - x_2)^2 = 2N$, the expectation value of Term 3–(Term 2)/2 $= 0$.

If there are n_f fitted values, one of which is the normalization and n_b bins in the histogram a modification is needed. The expectation value after a fit for a χ^2 with n_b bins and n_f fitted parameters is $n_b - n_f$. This occurs because, after fitting, the multidimensional normal distribution loses n_f variables.

There is an easy general way to handle this problem. When the simple χ^2 method is applied, the derivative of the χ^2 is in error because the change in the normalization of the particle density function is not taken into account. As for the toy model, including this term in the maximum likelihood approach eliminates the problem. This leads to a simple approach using a modified χ^2 analysis. Consider n_b bins and f fitting parameters p_j. Let $n_i(p_1, p_2, \cdots, p_f)$ be the expected number of events in bin i. The distribution of experimental events in each bin is taken as approximately normal. The total number of events in the histogram is not fixed. Choose the set n_i as the basis. The statistical error matrix is diagonal in this basis. Ignoring the 2π constants:

$$\ln \mathcal{L} = \sum_{i=1}^{n_b} -\frac{\ln n_i}{2} - \frac{(x_i - n_i)^2}{2n_i}.\tag{13.15}$$

$$\frac{d \ln \mathcal{L}}{dn_i} = \frac{x_i - n_i}{n_i} + \frac{1}{2n_i}\left[\left(\frac{(x_i - n_i)^2}{n_i}\right) - 1\right].\tag{13.16}$$

The expectation value for the term in square brackets is zero. Recall that the expectation refers to the average value over a number of repetitions of the experiment. It is x_i that changes with each experiment not the theoretical expectation, n_i. The expectation value of the term in square brackets will remain zero even if it is multiplied by a complicated function of the p_j fitting parameters. Ignoring this term leads to

$$\frac{\partial \ln \mathcal{L}}{\partial p_j} = \sum_{i=1}^{n_b} \left(\frac{x_i - n_i}{n_i}\right)\frac{\partial n_i}{\partial p_j}.\tag{13.17}$$

By expressing the n_i as the appropriate functions of the p_j, the error matrix can be written in terms of the p_j. However, the derivative of the inverse error matrix does not appear in the transform of Eq. 13.16. This result means that one can use a modified χ^2 approach. Use the usual χ^2, but, when derivatives are taken to find the χ^2 minimum, omit the derivatives of the inverse error matrix. The result is almost identical to the result from maximum likelihood. This is called the modified χ^2 minimum method and should generally be used in place of the regular χ^2 method.

In practice, since the differences are not precisely the expectation values for a given experiment, there is a small residual higher order effect, which causes no bias on the average.

However, if one of the standard minimizers is used, it will effectively take derivatives of the denominator. By using the minimizers to minimize the log of the likelihood, the correct parameter fits are obtained. However, minimizers such as MINUIT by default give proper error matrices for χ^2 fits. To get the errors correct, it is important to note that the likelihood has $\chi^2/2$ in the exponential. The minimizers have a variable to set how far above the minimum to go to set errors. The default value of this parameter is set for χ^2, so it is necessary to set that parameter to half of the χ^2 value.

The following formal theorem (Cramer 1946) justifies the use of the method of least squares for a histogram.

Suppose the p_i are such that for all points of a non-degenerate interval A in the s-dimensional space of the α_j, they satisfy

(a) $p_i > c^2 > 0$ for all i,
(b) p_i has continuous derivatives $\partial p_i/\partial \alpha_j$ and $\partial^2 p_i/\partial \alpha_j \, \partial \alpha_k$ for all $j, k = 1, s$,
(c) $D = \{\partial p_i/\partial \alpha_j\}$ for $i = 1, \ldots, r$; $j = 1, \ldots, s$ is a matrix of
      rank s.

Suppose the actual values of the parameters for the given experiment, are a vector, α_0, are an inner point of the interval A. Then the modified χ^2 minimum equations have one system of solutions and it is such that the solution, α^*, converges in probability to the actual α_{0j} as $n \to \infty$. [That is, $p(|\alpha^* - \alpha_0| > \delta)$ becomes $< \epsilon$ for any fixed δ, ϵ.] The value of χ^2 obtained from these equations is asymptotically distributed according to the χ^2 distribution with $r - s - 1$ degrees of freedom. Thus, we lose one degree of freedom per parameter we wish to estimate as per Fisher's lemma.

As an instructive example of the maximum likelihood method for histogram fitting, we will examine the maximum likelihood solution for a problem in which the least squares method can also apply. We have r categories and the probability of category i is p_i. We perform n experiments and v_i fall in category i. Then

$$\mathcal{L} = p_1^{v_1} p_2^{v_2} \cdots p_r^{v_r},$$

$$w = \log \mathcal{L} = \sum_{i=1}^{r} v_i \log p_i,$$

$$\frac{\partial w}{\partial \alpha} = \sum_{i=1}^{r} \frac{v_i}{p_i} \frac{\partial p_i}{\partial \alpha} = 0.$$

Note that we might argue that we should use

$$\mathcal{L} = \mathcal{L} \frac{n!}{v_1! v_2! \cdots v_r!},$$

since this represents the likelihood of the experimental number of events in each bin regardless of the order in which the events arrived. However, we see that this would not affect the derivative equation and only changes $\log \mathcal{L}$ by a constant.

Let us now compare the result above with the modified χ^2 minimum results. There (omitting the second term) we had

$$-\frac{1}{2}\frac{\partial \chi^2}{\partial \alpha} = \sum_{i=1}^{r}\left(\frac{v_i - np_i}{p_i}\right)\frac{\partial p_i}{\partial \alpha} = 0.$$

We now note that

$$-\sum_{i=1}^{r}\frac{np_i}{p_i}\frac{\partial p_i}{\partial \alpha} = -n\frac{\partial}{\partial \alpha}\sum_{i=1}^{r}p_i = -n\frac{\partial}{\partial \alpha}(1) = 0.$$

Hence, we see that indeed the two methods give the same result.

What is the relation of $w = \log \mathcal{L}$ and χ^2? They look very different, yet give the same equation for α^*. To relate them, we start with a somewhat different expression for $\mathcal{L}$. Suppose $v_i >> 1$ and $r >> 1$ and suppose for the moment we ignore the constraint

$$\sum_{i=1}^{r} v_i = n.$$

Then the probability that bin i has v_i events is approximately normal.

$$f(v_i) \cong \frac{1}{\sqrt{2\pi np_i}}\exp\left(-\frac{(v_i - np_i)^2}{2np_i}\right). \tag{13.18}$$

[We previously considered $f(x_j)$, the density function for the jth event. Now we are considering the probabilities of various numbers of events in a bin.] The likelihood function is now the product of the f_i for each bin.

$$w = \log \mathcal{L} = -\frac{r}{2}\log 2\pi n - \frac{1}{2}\sum_{i=1}^{r}\log p_i - \sum_{i=1}^{r}\frac{(v_i - np_i)^2}{2np_i}$$

$$= \frac{-r}{2}\log 2\pi n - \frac{1}{2}\sum_{i=1}^{r}\log p_i - \frac{1}{2}\chi^2. \tag{13.19}$$

Thus

$$w \cong -\tfrac{1}{2}\chi^2 + C.$$

where C is a constant for fixed p_i. Therefore, we see that for fixed p_i, $-2\log \mathcal{L} + 2C$ is asymptotically distributed in a χ^2 distribution with $r - 1$ degrees of freedom.

We have a remarkable situation now. We know from the main theorem on maximum likelihood estimates that asymptotically $\alpha - \alpha^*$ has a Gaussian density function. However, χ^2 as a whole is distributed according to a χ^2 distribution, which has much larger tails.

To see how this phenomenon comes about, we consider a specific example. Suppose we are taking samples from a normal distribution of known variance and wish to estimate the population mean.

$$\mathcal{L} = \prod_{i=1}^{n} \frac{1}{\sqrt{2\pi\sigma^2}} e^{-(x_i - m)^2/2\sigma^2},$$

$$\log \mathcal{L} = \text{constant} - \frac{1}{2\sigma^2} \sum_{i=1}^{n} (x_i - m)^2.$$

Let

$$Z = \sum_{i=1}^{n} (x_i - m)^2 = \sum_{i=1}^{n} (x_i - x_{AV} + x_{AV} - m)^2 \tag{13.20}$$

$$= \sum_{i=1}^{n} ((x_i - x_{AV})^2 + (x_{AV} - m)^2),$$

where

$$x_{AV} = \frac{1}{n} \sum_{i=1}^{n} x_i.$$

This last expression for Z results as the cross term sums to zero. We then have

$$Z = n(x_{AV} - m)^2 + \sum y_i^2,$$

where

$$y_i = x_i - x_{AV}.$$

Thus, $\tag{13.21}$

$$\mathcal{L} = e^{-\chi^2/2+C},$$

$$\chi^2 = \frac{(x_{AV} - m)^2}{\sigma^2/n} + \frac{1}{\sigma^2} \sum_{i=1}^{n} y_i^2.$$

Furthermore, x_{AV} and y_i are independent since the covariance of x_{AV} and y_i, i.e., $C(y_i, x_{AV})$, is zero:

$$C(y_i, x_{AV}) = \mathbb{E}\{(y_i - \mathbb{E}\{y_i\})(x_{AV} - \mathbb{E}\{x_{AV}\})\}. \tag{13.22}$$

But,

Fig. 13.1 Plot of log $\mathcal{L}$ vs α^*. The width $\Delta\alpha^*$ is estimated by observing where log $\mathcal{L}$ has changed by 1/2 from its maximum value.

$$\mathbb{E}\{x_{AV}\} = m, \ \mathbb{E}\{y_i\} = \mathbb{E}\{x_i - x_{AV}\} = m - m = 0,$$

$$C(y_i, \ x_{AV}) = \mathbb{E}\{y_i(x_{AV} - m)\} = \mathbb{E}\{y_i x_{AV}\},$$

$$\mathbb{E}\{x_i x_{AV} - (x_{AV})^2\} = \mathbb{E}\left\{\frac{1}{n}\sum_{j=1}^{n} x_i x_j - \frac{1}{n^2}\sum_{j,k} x_j x_k\right\}$$

$$= \frac{n-1}{n}m^2 + \frac{(\sigma^2 + m^2)}{n} - \frac{1}{n^2}n(n-1)m^2$$

$$- \frac{n}{n^2}(\sigma^2 + m^2) = 0.$$

(13.23)

We see that χ^2 breaks up into two independent terms. The first term is the sufficient statistic and the second term is what is left over. As a function of x_{AV}, we see $\mathcal{L}$ is Gaussian and the remaining part is χ^2 in $n - 1$ degrees of freedom (the y_i are not all independent among themselves). The expectation value of the first term is 1 and the maximum likelihood estimate of the mean is $m^* = x_{AV}$. Hence, the value of χ^2 minimum is determined mainly by the second term. The variation of $\mathcal{L}$ caused by change of x_{AV} and, hence, the variance of x_{AV} is determined by the first term only. Thus, the value of χ^2 at the minimum is decoupled from the width of the distribution as function of x_{AV}. This result generalizes for a multiparameter problem. It is also quite similar to the result quoted in the last chapter for a likelihood ratio.

From the above results, we see that by plotting the experimental log $\mathcal{L}$ as a function of the estimated parameter α^*, we can estimate the width. $\sigma \cong \Delta\alpha^*$, where $\Delta\alpha^*$ is the change in α^* from the maximum value required to change log $\mathcal{L}$ by $\frac{1}{2}$. See Fig. 13.1.

As an aside, we note that this result implies that if we take the weighted mean of n measurements, say n measurements of the lifetime of a given nuclear species, then $\alpha^* - \alpha$ is distributed in a Gaussian distribution, but χ^2 of the mean is distributed in a χ^2 distribution with $n - 1$ degrees of freedom. This has larger tails than a Gaussian distribution. Thus, when looking at several experimental measurements of a single parameter, we may find larger discrepancies in fact consistent than we might think at first assuming a Gaussian distribution.

13.3   Further Considerations in Fitting Histograms

Let us turn again to the problem of comparing a curve to a histogram with r bins and ν_i observed events in bin i.

$$\mathcal{L} = p_1^{\nu_1} p_2^{\nu_2} \cdots p_r^{\nu_r}.$$

If the total number of events $\nu = \sum_{i=1}^{r} \nu_i$ is fixed, then this would be used unmodified with the constraint $\sum_{i=1}^{r} p_i = 1$. Suppose we define χ^2 by

$$\chi^2 = \sum_{i=1}^{r} \frac{(\nu_i - \nu p_i)^2}{\nu p_i}. \tag{13.24}$$

One degree of freedom is lost because the sum of the bins is fixed. It can be shown that

$$\mathbb{E}\{\chi^2\} = r - 1; \quad \text{var}\{\chi^2\} = 2(r - 1) + \frac{1}{n} \sum_{i=1}^{r} \left(\frac{1}{p_i} - r^2 - 2r + 2 \right). \tag{13.25}$$

The second term in the variance expression reflects the fact that, at finite n, the χ^2 defined here only approximately follows a χ^2-distribution.

If events are grouped into histograms and all events assumed to be at the midpoint of each bin for calculating moments of the distribution, a bias is introduced. It is better to use the actual values of each event. Suppose the bins are all of uniform width h. Let μ_i be the ith central moment using the actual values for each event and $\overline{\mu}_i$ be the corresponding central moment using the bin midpoint values for each event. Then the corrections are approximately given by

$$\mu_2 = \overline{\mu}_2 - \frac{1}{12} h^2,$$

$$\mu_3 = \overline{\mu}_3, \tag{13.26}$$

$$\mu_4 = \overline{\mu}_4 - \frac{1}{2} \overline{\mu}_2 h^2 + \frac{7}{240} h^4.$$

These are Sheppard's corrections for grouping. If these corrections are large, or if the function varies significantly over a bin, it is likely that using midpoint values for any calculations, not just moments, may lead to serious bias.

Suppose, in a histogram, that the total number of events v is not fixed, but is assumed to vary according to a Poisson distribution, with n the expected total number of events. Then

$$\mathcal{L} = \frac{e^{-n} n^v}{v!} p_1^{v_1} p_2^{v_2} \cdots p_r^{v_r}.$$

$$w = \log \mathcal{L} = \sum_{i=1}^{r} v_i \log p_i + v \log n - n = \sum_{i=1}^{r} v_i \log(n p_i) - n.$$

We ignored $v!$ in the last relation. It is an additive constant in w, independent of the parameters, p_i. Note $n_i \equiv n p_i$ is the expected number of events in bin i.

$$w = \sum_{i=1}^{r} (v_i \log n_i - n_i), \tag{13.27}$$

where there are now no constraints on the n_i. The number in each bin varies according to a Poisson distribution. This is known as the extended maximum likelihood method.

By adding terms that don't depend on the parameters n_i, an equivalent form, often useful for analysis is

$$w = \sum_{i=1}^{r} \left[(v_i - n_i) + v_i \log(n_i / v_i) \right]. \tag{13.28}$$

13.4 Estimation of a Correlation Coefficient

A special problem arises when one tries to estimate a correlation coefficient for a multivariable problem. The density function for a correlation coefficient is complicated, asymmetric, and broad. This leads to large errors in the estimates. Define the sample correlation coefficient by

$$r = \frac{\sum_{i=1}^{n} (x_i - x_{AV})(y_i - y_{AV})}{\sqrt{(\sum_{j=1}^{n} (x_j - x_{AV})^2)(\sum_{j=1}^{n} (y_j - y_{AV})^2)}}, \tag{13.29}$$

which is bounded by $-1 \leq r \leq 1$.

Fisher (1990) suggests using the experimental variable

$$z = \tanh^{-1} r = \frac{1}{2} \log \frac{1+r}{1-r} \tag{13.30}$$

as an estimate of ζ,

$$\zeta = \tanh^{-1} \rho = \frac{1}{2} \log \frac{1+\rho}{1-\rho}, \tag{13.31}$$

where ρ is the correlation coefficient. One finds the expectation value and variance of z to be

$$\mathbb{E}\{z\} = \frac{1}{2} \log \frac{1+\rho}{1-\rho} + \frac{\rho}{2(n-1)}. \tag{13.32}$$

$$V\{z\} = \frac{1}{n-3}. \tag{13.33}$$

The bias is seen to fall off with $1/n$, and the variance is more favorable than that for r.

In this chapter, we have introduced methods for estimating parameters from data. When appropriate we have quoted the detailed theorems justifying the procedures and indicated the conditions required for their validity. The maximum likelihood method and the closely related least squares methods are two of the most important methods in practice.

13.5   Worked Problems

WP13.1 Suppose we have a counter setup which counts pulses from a long-lived radioactive source for 1 sec. (Ignore the change in the expected rate over the measurement time. We assume this is a very long-lived isotope.)
We obtain n counts. What is the maximum likelihood estimate for the mean number of counts and the variance of this estimate?

Answer:

We have made a measurement and gotten n. We believe the number to be distributed in a Poisson distribution and ask for the maximum likelihood estimate of the mean number of counts.

$$\mathcal{L} = \frac{e^{-\lambda}\lambda^n}{n!},$$
$$w = \log \mathcal{L} = -\lambda + n \log \lambda - \log n!,$$
$$\frac{\partial w}{\partial \lambda} = -1 + \frac{n}{\lambda} = 0 \text{ at the maximum,} \tag{13.34}$$
$$\lambda^* = n.$$

The variance of λ^* asymptotically is given by

$$\text{variance}(\lambda^*) = \left(\mathbb{E}\left\{-\frac{\partial^2 w}{\partial \lambda^2}\right\}\right)^{-1} = \left(\mathbb{E}\left\{\frac{n}{\lambda^2}\right\}\right)^{-1}.$$

Now λ^2 is a constant as far as the expectation value is concerned and can be pulled out. Furthermore, $\mathbb{E}\{n\} = \lambda$ for a Poisson distribution. Hence,

$$\text{variance}(\lambda^*) = \lambda \cong n. \tag{13.35}$$

WP13.2 Suppose we have made a series of n measurements of the same quantity $x_1, x_2, \ldots, x_n$ in which each measurement has a known variance $\sigma_1, \sigma_2, \ldots, \sigma_n$, different for each measurement. Assume each measurement distributed with normal distribution. Find the maximum likelihood estimate of the   value of this quantity and the variance of this estimate.

Answer:

We have seen that the χ^2 test for the mean of a normal distribution with known variance is obtained by the maximum likelihood method.

Our present problem is a slight generalization of this. We have made a series of n measurements $x_1, x_2, \ldots, x_n$ in which each measurement has known variance $\sigma_1, \sigma_2, \ldots, \sigma_n$. Here the variances can differ for each measurement. This is approximately the situation when we are trying to combine different experimental measurements of, say, a particle lifetime.

$$\mathcal{L} = \prod_{i=1}^{n} \frac{1}{\sqrt{2\pi\sigma_i^2}} \exp\left(-\frac{(x_i - m)^2}{2\sigma_i^2}\right),$$

$$w = \log \mathcal{L} = \sum_{i=1}^{n}\left(-\frac{1}{2}\log 2\pi - \frac{1}{2}\log \sigma_i^2 - \frac{(x_i - m)^2}{2\sigma_i^2}\right),$$

$$\frac{\partial w}{\partial m} = \sum_{i=1}^{n} \frac{x_i - m}{\sigma_i^2} = 0 \text{ at the maximum,}$$

$$m^* = \frac{\sum_{i=1}^{n} x_i/\sigma_i^2}{\sum_{i=1}^{n} 1/\sigma_i^2},$$

$$\text{variance}(m^*) \to \left(\mathbb{E}\left\{\frac{-\partial^2 w}{\partial m^2}\right\}\right)^{-1} = \left(\mathbb{E}\left\{\sum_{i=1}^{n}\frac{1}{\sigma_i^2}\right\}\right)^{-1},$$

$$\frac{1}{\sigma_{m^*}^2} = \sum_{i=1}^{n}\frac{1}{\sigma_i^2}.$$

This reproduces the results obtained in Worked Problem 3.2. Here we have made the additional assumption that each measurement is normal and obtain the additional result that m^* is approximately distributed in a normal distribution

WP13.3 Suppose we take n samples from the *same* normal distribution but we do not know the mean or variance of the distribution. Use the maximum likelihood method to estimate the mean and variance of the normal distribution. (You are not required to find the variances of these estimates.)

Answer:

We have taken n samples from the same normal distribution, but know neither the mean nor the variance. For example, suppose we are measuring the range of a set of α particles of the same energy coming from the decay of a single state of a particular kind of radioactive nucleus. We wish to measure the range and the variance (straggling) and we make the assumption that the distribution is at least approximately normal (see, however, the discussion of straggling in Chap. 11).

$$\mathcal{L} = \frac{1}{(2\pi\sigma^2)^{n/2}} \exp\left(-\frac{1}{2\sigma^2} \sum (x_i - m)^2\right),$$

$$w = \log \mathcal{L} = \frac{-n}{2}(\log 2\pi\sigma^2) - \sum_{i=1}^{n} \frac{1}{2\sigma^2}(x_i - m)^2,$$

$$\frac{\partial w}{\partial m} = \sum_{i=1}^{n} \frac{(x_i - m)}{\sigma^2} = 0 \text{ at the maximum,}$$

$$m^* = \frac{1}{n} \sum_{i=1}^{n} x_i = x_{\text{AV}},$$

$$\frac{\partial w}{\partial \sigma^2} = \frac{-n}{2\sigma^2} + \frac{1}{2\sigma^4} \sum_{i=1}^{n} (x_i - m)^2 = 0 \text{ at the maximum,}$$

$$\sigma^{2*} = \frac{1}{n} \sum_{i=1}^{n} (x_i - m^*)^2.$$

We note that this is the most probable result, but it is a biased result. Using $1/(n-1)$ instead of $1/n$ would make it unbiased. The maximum likelihood method can produce a biased estimate. It does have to be asymptotically unbiased.

13.6 Exercises

13.1 Suppose we have a visible detector such as a bubble chamber with a fast beam of short-lived particles of known energy passing through it. We might imagine, for instance, that we have a beam of 5.0 GeV Σ^- particles. Some of the particles decay in the chamber and some pass through the chamber without decaying. We wish to measure the lifetime by finding the distribution of decay lengths in the detector. The overall probability of decaying at a distance between x and $x + dx$ from the beginning of the chamber is proportional to $e^{-x/L}$ ($x > 0$), where L is the mean length for decay and is the parameter we wish to estimate. The length of the chamber is d. The decay point of a track that decays in the chamber is marked by a kink in the track. We measure the position x_i of each decay point. Estimate L and its variance using a maximum likelihood method for the cases that:

(a) we have other non-decaying tracks passing through the detector besides the wanted particles, and hence have no information on the number of particles not decaying;
(b) we do know how many of the wanted particles pass through the detector and do not decay in it, either by having a pure beam to begin with or by having counters outside the chamber capable of tagging the desired kind of particle.

For finding the variance of the estimates in practice, it may be more convenient to use $\partial^2 w/\partial L^2$ from the experimental values rather than using the expectation values. This is often done in practice. Although not required here, this problem can be profitably treated by the Bartlett S function modification of the maximum likelihood method which will be discussed in Chap. 17. In fact, the method originally was developed specifically for this problem.

13.2 Imagine that we measure n pairs of x, y values. Suppose y is a normally distributed variable with constant variance σ^2 but with a mean that depends linearly on x.

$$m = \alpha + \beta(x - \bar{x}),$$

where $\bar{x}$ is the sample mean. x does not have to be a random variable. We might fix x at x_1 and sample y. Then we might fix x at x_2 and sample y again, etc., σ^2 is not known. Here x might be the temperature of a sample and y be the heat capacity. Show that the maximum likelihood estimate for the parameters α, β, σ^2 are

$$\alpha^\star = \frac{1}{n} \sum_{i=1}^{n} y_i,$$

$$\beta^\star = \lambda_{12}/s_1^2,$$

where

$$s_1^2 = \frac{1}{n} \sum_{\nu=1}^{n} (x_\nu - \bar{x})^2, \lambda_{12} = \frac{1}{n} \sum_{\nu=1}^{n} (x_\nu - \bar{x})(y_\nu - \bar{y}),$$

and

$$\sigma^{2\star} = s_2^2(1 - r^2) = \frac{1}{n} \sum_{\nu=1}^{n} (y_\nu - \alpha^\star - \beta^\star(x_\nu - \bar{x}))^2,$$

where

$$s_2^2 = \frac{1}{n} \sum_{\nu=1}^{n} (y_\nu - \overline{y})^2, r = \frac{\lambda_{12}}{s_1 s_2}.$$

(r is the sample correlation coefficient.)

13.3 A set of n thin wire proportional chambers uniformly spaced at positions along the x axis has made a series of measurements of track position, i.e., x, y values for a straight track. We wish to estimate $\tan\theta$ in the two-dimensional projection. The spacing between individual chambers is Δx. Find an estimate for $\tan\theta$ and its variance as a function of n. Assume the measurement error is normal and the same at each chamber. It is to be estimated from the data. We are particularly interested in how the variance changes with Δx and with n. We ignore multiple scattering errors, errors in x, and systematic errors, i.e., distortions. How would the results change if we knew the measurement error of the chambers in advance? Hint:

$$\sum_{\nu=1}^{n} \nu = \frac{1}{2} n(n+1),$$

$$\sum_{\nu=1}^{n} \nu^2 = \frac{n}{6}(n+1)(2n+1).$$

13.4 For the toy model in Sect. 13.2, we had a naive solution for the estimate of the number of events in (13.7) and a solution in which the variance was defined as x_i rather than the theoretically predicted x in (13.10). Show that for this toy model, if ϵ is defined by $x_2 = x_1(1 + \epsilon)$, then the two solutions differ only by a term of order ϵ^2.

References

Cramer H (1946) Mathematical methods of statistics. Princeton University Press, Princeton

Fisher R (1990) Statistical methods, experimental design and scientific inference, a re-issue of statistical methods for research workers. The design of experiments and statistical methods and scientific inference. Oxford University Press, Oxford

Chapter 14
Inverse Probability; Confidence Limits

Abstract We discuss Bayes' theorem and problems of A Priori probabilities. Confidence intervals are described and suggestions and limitations for improvements in the selection of a confidence interval are discussed. The handling of confidence intervals with partial background-signal separation is discussed. Finally, the difficult problem of deciding when a signal is significant, rather than an artifice is discussed.

14.1 Bayes Theorem

Suppose we have a set of a great many systems of k mutually exclusive kinds, i.e., systems of kinds $H_1, H_2, \ldots, H_k$. Suppose further that we randomly pick a system and perform an experiment on it getting the result A. What is the probability that we have a system of kind ℓ? This probability is the conditional probability that given A, we have H_ℓ, i.e., $P\{H_\ell | A\}$. But we know that this is the probability of getting H_ℓ and A divided by the overall probability of getting A. Thus

$$P\{H_\ell | A\} = \frac{P\{AH_\ell\}}{P\{A\}}. \tag{14.1}$$

This follows easily by noting that $P\{A\}P\{H_\ell | A\}$ is just the probability of both A and H_ℓ. Similarly

$$P\{A\} = \sum_{j=1}^{k} P\{A | H_j\} P\{H_j\},$$
$$P\{AH_\ell\} = P\{A | H_\ell\} P\{H_\ell\}. \tag{14.2}$$

Note the difference between $P\{A | H_\ell\}$, the probability of getting A if H_ℓ is chosen (the direct probability) and the previous $P\{H_\ell | A\}$, the probability of having H_ℓ if result A is obtained (the inverse probability).

© Springer Nature Switzerland AG 2020
B. P. Roe, *Probability and Statistics in the Physical Sciences*,
Undergraduate Texts in Physics,
https://doi.org/10.1007/978-3-030-53694-7_14

Hence, we finally obtain

$$P\{H_\ell|A\} = \frac{P\{A|H_\ell\}P\{H_\ell\}}{\sum_{j=1}^{k} P\{A|H_j\}\, P\{H_j\}}. \tag{14.3}$$

This result is known as Bayes theorem. If the H_j are called causes, then this gives the probability of causes. $P\{H_j\}$ is known as the a priori probability of H_j. The generalization to continuous variables is trivial and is illustrated in an example below.

There are many occasions in which Bayes theorem are used. These often involve examples in which we have some a priori knowledge of a distribution. For example, suppose we have a beam of particles, say electrons, going in the z direction. Suppose we know the velocity distribution of this beam to be normal $(v_0,\ \sigma_0)$. Now we perform an experiment and measure the velocity of a specific particle, obtaining an answer v_1. We assume our measurement errors to be normal with standard deviation σ_1. What is our best estimate of the velocity of the measured particle?

In this example, we can apply Bayes theorem to obtain our best estimate of the velocity. We modify the above trivially because v is a continuous variable here. Then

$$f(v|v_1)\, dv = \frac{\left(\frac{1}{\sqrt{2\pi\sigma_1^2}}\exp\left(-\frac{(v-v_1)^2}{2\sigma_1^2}\right)\right)\left(\frac{1}{\sqrt{2\pi\sigma_0^2}}\exp\left(-\frac{(v-v_0)^2}{2\sigma_0^2}\right)\right)dv}{\int_{-\infty}^{\infty}\left(\frac{1}{\sqrt{2\pi\sigma_1^2}}\exp\left(-\frac{(v-v_1)^2}{2\sigma_1^2}\right)\right)\left(\frac{1}{\sqrt{2\pi\sigma_0^2}}\exp\left(-\frac{(v-v_0)^2}{2\sigma_0^2}\right)\right)dv}. \tag{14.4}$$

This corresponds to a normal distribution for v, but not centered at v_1 (Fig. 14.1).

For another example, consider a blood test for a certain disease that gives a positive result 97% of the time if the disease is present and gives a false positive result 0.4% of the time when the disease is not present. Suppose 0.5% of the population actually has the disease. We can ask what the probability is that a person actually has the disease if that person's test is positive. Let H be the probability of having the disease and A be the probability of testing positive. We know

$$P\{A|H\} = 0.97,\ P\{H\} = 0.005,\ P\{A \mid \text{not } H\} = 0.004,$$

$$P\{H|A\} = \frac{0.97 \times 0.005}{0.97 \times 0.005 + 0.004 \times 0.995} = 0.55.$$

This means that the results giving a positive test will get 97% of people really having the disease, but $100 - 55\% = 45\%$ of the positive tests will be for people not having the disease.

Fig. 14.1 Illustration of the use of Bayes theorem in estimating the velocity of a particle beam. The dashed line illustrates the combined density function $f(v|v_1)$

14.2 The Problem of A Priori Probability

Perhaps no other part of probability study has sustained more controversy than the discussions about the use and misuse of Bayes theorem. It is surprising that a subject that would seem to be amenable to mathematics would lead to such passionate discussions. The problem with applications is that an a priori initial probability needs to be specified.

Bayes theorem is clearly true as proven above but caution is needed to apply it in practice with an arbitrary a priori probability $P\{H_\ell\}$. In the examples above, we know the a priori distributions. *In many applications made of Bayes theorem in the literature, one sets $P\{H_\ell\}$ equal to a constant independent of ℓ.* This choice is not invariant under the transformation of variables. There are a few other standard a priori distributions used, but all of them are arbitrary.

In addition, we often use it when the result is not a matter of probability. Suppose we measure the charge on an electron, e. What does probability mean here? We do not have an ensemble of universes each with a different value for e. We have available only the universe in which we live. Here the answer has one value and it is not strictly a matter of probability. What we are doing is attempting to use probability as a measure of our state of knowledge of a quantity. This can be very useful, but it also can lead us into some subtle traps and it becomes important to use language very carefully.

Parratt (1961) called the arbitrary a priori assumption the "hypothesis of desperation". In desperation, we choose $P\{H_\ell\}$ equals a constant. However, it has become clear that in some situations it is quite appropriate.

For reporting data, we can sometimes avoid the use of Bayes theorem. The frequentist method avoids the arbitrariness of Bayes methods. Suppose an experiment is trying to determine the value of a parameter μ and obtains a value x. Frequentists ask that given x, what is the range of μ such that, the probability of getting a result x, is greater than, say 10%. We will discuss this method in the next chapter.

We will shortly consider an instance, however, even for reporting data, in which we use Bayes theorem arguments and then interpret the result in a frequentist test. In recent years there has been a realization that often, even with experimental data, there is only partial knowledge about biases and correlations with the data and uncertainties in the theoretical interpretation. Here the prior is not arbitrary and the use of Bayes theorem has dramatically increased. These models, *marginal models*, average over the uncertainties in the priors. In this sense, they are different from a repetition of the experiment which uses only the real value, not the average.

Suppose we are faced with deciding whether to accept a hypothesis or not. For example, if a result is true, we may wish to do a new experiment to examine it further. Should we invest the time and resources to perform this new experiment? Here, we wish to use all of the information we have, qualitative and quantitative, to make our decision. In this situation, it is natural to use our best guess as to the a priori probability in addition to using the data results to aid our decision. Use of Bayesian methods for decision-making is appropriate. Even after reporting, frequentist limits for the results from a given experiment, it is often appropriate to discuss the conclusions from the data using Bayesian methods.

In Bayesian inference about the parameter p of a binomial process, if the prior p.d.f. is a beta distribution $f(p|\alpha, \beta)$ then the observation of r successes out of n trials gives a posterior beta distribution $f(p|r + \alpha, n - r + \beta)$. We will return to Bayesian questions in Sects. 14.4–14.5, but we will first discuss the frequentist method of setting confidence intervals.

14.3   Confidence Intervals and Their Interpretation

We will begin this section by reviewing some of the methods we have discussed for setting an error on a measured quantity. Let us consider a specific example. Suppose we search for a very rare particle decay in a long run. For example, we might search for $k^+ \rightarrow \pi^+\pi^+\pi^-\gamma$ in a run involving many thousands of k^+ decays. Imagine we find three events. We wish to quote an error on our experimental rate. We know that the number of events observed will have a Poisson distribution. Suppose we ask,

"What is the error?" Consider different estimates of the error. What is the probability the mean was 10 events?

If we make the assumption that the observed number of events is the mean, i.e., that $\lambda = 3$, and we then assume a Poisson distribution, the probability of 10 or more events is $P = 0.001$. However, there is a problem with this assumption. σ should be proportional to $\sqrt{n}$ expected not $\sqrt{n}$ observed.

Suppose, instead of the above, we ask a different question, "I have a physical theory that predicts $\lambda = 10$. What are the chances of obtaining three or fewer events in this experiment?"

$$P = \sum_{n=0}^{3} \frac{e^{-\lambda}\lambda^n}{n!} = e^{-10}\left(\frac{10^3}{3!} + \frac{10^2}{2!} + \frac{10}{1!} + 1\right)$$

$$= \frac{e^{-10}}{6}(1000 + 300 + 60 + 6) \approx 0.01.$$

(14.5)

This last question is far more precise than the previous one. It does not require arbitrary assumptions (the mean is the observed number.) By properly phrasing our questions, we can often avoid these vague procedures.

We now turn to the discussion of confidence levels. Suppose we are measuring a physical parameter α. α^* is the result of our measurement for α. Let $g(\alpha^*|\alpha)$ be the density function for the estimate α^*, for fixed parameter value α. For each α, we define (γ_1, γ_2) such that the probability of α^* falling outside these limits is ϵ

$$\int_{\gamma_1}^{\gamma_2} g(\alpha^*|\alpha)\,d\alpha^* = 1 - \epsilon.$$

(14.6)

Clearly, γ_1 and γ_2 are not unique. We can change them at will to accept more of one tail of the distribution and less of the other while keeping ϵ constant. For a given choice of percentages of each tail, γ_1 and γ_2 are functions of α. To start with, we will assume that we choose symmetric tails with probability $\epsilon/2$ in each tail.

In Fig. 14.2, we have a plot in (α, α^*) space. The two curved lines represent γ_1 and γ_2 as functions of α. The dashed line is a line of constant α and we know that for α fixed, the probability of having our sample estimate α^* fall between the lines γ_1 and γ_2 is $1 - \epsilon$. Thus, for any given α, the probability that our result α^* falls in the shaded region D is $1 - \epsilon$. This is the central point in our argument. Please look at Fig. 14.2. If the actual α (horizontal line) intersects the experimental result (vertical line) inside the shaded region, this is in the $1 - \epsilon$ probability region, whatever the value of α. Suppose now we perform an experiment obtaining the result α^*_{exp}, which cuts the line γ_1 and γ_2 at $\alpha = c_1$ and $\alpha = c_2$, respectively. If now we estimate that α is between c_1 and c_2, i.e., that our result has fallen in the shaded region D, we see that we are right $1 - \epsilon$ of the time. That is, if $1 - \epsilon = 0.9$, and we do many trials finding c_1, c_2 for each *trial*, then for 90% of the trials, α will be between c_1 and c_2 found for that trial. For a given trial, the interval c_1 to c_2 is then called a $100 \times (1 - \epsilon)$ percent confidence interval. The argument is easily extended to

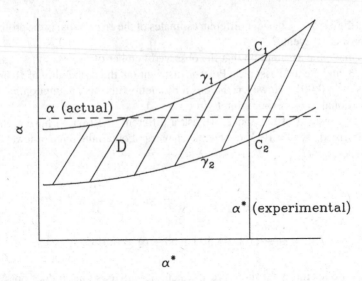

Fig. 14.2 The confidence level method. The curves γ_1 and γ_2 represent fixed values for the probability of the experimental estimate, α^*, of the theoretical parameter, α, being farther from the actual value than indicated and in the same direction

Light   Test   Standard sheet Photocell
        polaroid sheet

Fig. 14.3 Experimental setup for measuring the polarization efficiency of a test polaroid sheet

problems involving several parameters. The use of confidence levels calculated as we have indicated is known as the frequentist approach to setting limits. The confidence area D is called the confidence belt. This concept originated with Neyman (1937), with further work by E. S. Pearson and by students of Lehman (1986). The above statements on confidence intervals are precise and, as stated, do not take a priori probabilities into account. Occasionally, one does have some a priori knowledge and can use this to obtain improved estimates. For example, suppose we are measuring the polarizability of a sheet of polaroid. We have a standard sheet with polarizability P essentially 1.

Using the apparatus shown in Fig. 14.3, we measure the light transmitted as a function of θ, the angle between the direction of polarization of our standard and our test polaroid. In Fig. 14.4, the crosses are our experimental results and the curve is a best fit curve to those results using the function

$$I = I_0(1 + \alpha \cos 2\theta),$$

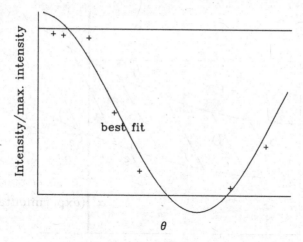

Fig. 14.4 Resulting plot of data and experimental fit

Fig. 14.5 Example of result in which entire confidence interval lies in a forbidden region

where α is our parameter. The best fit curve here corresponds to a value of α slightly greater than 1. This is a physical impossibility since the intensity would have to be less than zero at the some point. To dramatize the problem, suppose the result is such that the entire confidence interval lies in this forbidden region $\alpha > 1$, if symmetric tails are chosen. This is shown in Fig. 14.5.

This does not happen often, certainly less than $1 - \epsilon$ of the time. However, when it does happen we know that the values in the confidence interval are wrong. We will address this defect shortly. In general, what the interval means is that you are confident that for $1 - \epsilon$ of the trials α is contained within the limits (c_2, c_1). For a specific trial,

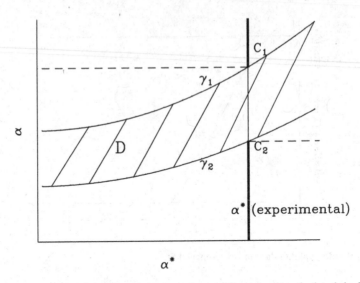

Fig. 14.6 Illustration of the difference between the confidence level method and the Bayesian credible region. The confidence level method uses the probability integrated over the horizontal dashed lines shown and the Bayesian credible region integrates over the heavy vertical lines shown

you may have other information and not be $1 - \epsilon$ confident that α is within these limits. The statements above, sum up the results of our experimental knowledge precisely. A discussion of some of these same problems using non-symmetric tails is given in Sect. 14.5.

Consider Fig. 14.6. For our method, ϵ refers to the probability for *fixed* α of α^* falling outside the interval D. Thus, it is the probability integrated over the dashed horizontal lines in the figure. With a Bayesian approach, we consider the physical parameter α to be a variable. We obtain ϵ by integrating over α with α^* fixed. This is the path denoted by the solid vertical lines in the figure and the accepted region is called the credible region.

In general, these results need not be the same. However, for a normal distribution with known σ and a priori probability constant from $-\infty \leq m \leq \infty$, we will obtain the same answer. This occurs because $f(x|m) \propto f(m|x)$ under the hypothesis of desperation, and f is symmetric in x and m. We can also show that we obtain the same results for a Poisson distribution, if the a priori probability of the mean m is constant for $0 \leq m \leq \infty$. In general, as shown in Fig. 14.6, the confidence levels, ϵ, are obtained in a different manner and would not be expected to be the same.

In practice, we sometimes see people use confidence intervals in a manner different from the above. An interval (c_2, c_1) of α is set in advance of the experiment. The experiment is then performed, and using the result, a determination is made of the confidence level to which c_1 and c_2 correspond. Here the confidence interval is simply used as a rule. It is often useful. However, the confidence interval obtained in this way cannot be interpreted as a probability in a simple manner.

Table 14.1 Estimates of the mean for 10 measurements on a normal distribution

Case	σ	Parameter	Estimate	Estimate of variance of estimate
1	Known	m	x_{AV}	$\sigma^2/10$
2	Unknown	m	x_{AV}	$s^2/10$

Let us now look at some examples of confidence intervals. Suppose we make 10 measurements of a normally distributed variable. We consider two cases. In the first case, the variance of the distribution is known in advance and in the second case, it is unknown. See Table 14.1.

s^2 is given by the equation below

$$s^2 = \frac{1}{n-1}\sum_{i=1}^{n}(x_i - x_{AV})^2 = \frac{1}{9}\sum_{i=1}^{10}(x_i - x_{AV})^2.$$

In case 2, $(x_{AV} - m)/(s/\sqrt{10})$ is called a pivotal quantity. This means it is a quantity that is a sufficient statistic, i.e., one depending on the parameter wanted and no other, and in addition, it has a fixed distribution regardless of the outcome of the experiment.

Suppose one asks for a symmetric 95% confidence level for m. For case 1, $(x_{AV} - m)/(\sigma/\sqrt{10})$ is normal (0, 1) and the interval is $x_{AV} \pm 1.96\sigma/\sqrt{10}$. For case 2, $(x_{AV} - m)/(s/\sqrt{10})$ has a Student's distribution with nine degrees of freedom and the interval is $x_{AV} \pm 2.26s/\sqrt{10}$. We see the interval is smaller if σ is known. The meaning of the interval for case 2 is that if we make lots of sets of 10 measurements of normally distributed variables, then 95% of the time x_{AV} will be in the interval

$$|(x_{AV} - m)/(s/\sqrt{10})| \leq 2.26$$

14.4 Use of Confidence Intervals for Discrete Distributions

We have discussed an example of a discrete distribution in the preceding section. However, the discreteness of the distribution can introduce difficulties into the use of confidence intervals, since only discrete values are obtainable on one axis.

Consider a counting problem in which one counts events over a fixed time interval, obtaining n events. Suppose the distribution of the number of counts is a Poisson distribution and one wishes to get information on the mean λ of the distribution. In particular, imagine that one wishes to obtain an upper limit of λ. For a Poisson distribution with mean 2.3, there is a 10% chance that one will obtain 0 events. If the mean is greater than 2.3, there is less than a 10% chance that 0 events will be obtained. For a Poisson distribution with mean 3.89 there is a 10% chance that there will be 0 or 1 event obtained.

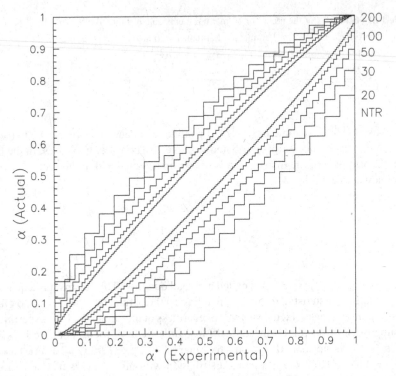

Fig. 14.7 The symmetric 95% confidence interval for a binomial distribution

For a confidence interval in a continuous distribution, the interval is determined such that for any value of the parameter in that region, the chance of being in the region is 90%. For a discrete distribution, the confidence region is chosen to be at least 90%, not precisely 90%. There are step function drops in the percentage of the time the limit is right. This percentage is known as the coverage. As the mean increases past 2.3, the coverage increases from 90% because the probability of zero events decreases. Finally 3.89 is reached and the coverage again drops discontinuously to 90%, etc. Hence, except for a set of discrete points, the coverage is always greater than 90%.

Figure 14.7 shows a 95% confidence interval graph for a binomial distribution. Reference (DeLury and Chung 1950), gives some confidence interval results for sampling of defectives from a finite population, that is, for the hypergeometric distribution.

Next, suppose that events are detected with an efficiency ϵ that is less than one. Two cases will be considered. In the first case, the efficiency varies randomly from trial to trial, owing to uncontrollable environmental effects. Assume we have established that the efficiency varies with known mean and variance. Often the precise distribution is unknown. It is inconvenient to use a normal distribution, since that would extend to negative efficiencies. Empirically, a useful distribution to chose is the gamma distribution, described in Sect. 5.2. The gamma distribution is given by

$$f(y) = \frac{\mu e^{-\mu y}(\mu y)^{\alpha-1}}{(\alpha-1)!}, \qquad y \geq 0, \tag{14.7}$$

and has mean α/μ and variance α/μ^2. We can choose α and μ to give any desired mean and variance. If we have a Poisson distribution with mean θ, where the probability of θ is given by the gamma distribution, then the overall probability of n events is given by the negative binomial distribution introduced in Exercise 5.1

$$P(r, m) = \frac{(r-1)!}{(m-1)!(r-m)!} p^m (1-p)^{r-m}, \tag{14.8}$$

with, here

$$m = \alpha; \quad r = n + \alpha; \quad p = \frac{\mu}{\mu+1}. \tag{14.9}$$

The mean of the negative binomial distribution is m/p and the variance is $m(1 - p)/p^2$. [For non-integral x, we define $x! \equiv \Gamma(x + 1) \equiv \int_0^\infty e^{-t} t^x dt$.] The proof that one obtains a negative binomial distribution is left to the Exercises.

In the second case, the efficiency is fixed and measured, and it has been established that the measurement of efficiency is distributed normally with known variance. Here, there is a fixed efficiency and one cannot fold the two distributions together as above without introducing a Bayesian hypothesis. However, we can ask, for any values of the measured number of events and measured efficiency, what the joint probability is that one obtains the combined result or less for a given real efficiency and real mean. This can then replace the previous limit. This is, then, a 90% C.L. for the joint result of measured efficiency and measured mean, not just the measured mean, i.e., there is now a two-dimensional confidence region. For a given measured mean and efficiency, the 90% C.L. for the true mean value of λ is obtained by taking the extreme values of the mean within that region. (In practice, the Bayesian assumption is often made, and this second case is treated just as the first.)

Note that these considerations also apply to the situation where one is obtaining a Poisson distribution in the presence of background, where $\theta = \lambda + b$, where λ is fixed and b is known with an error corresponding to case 1 or 2.

It has been noted Cousins and Highland (1992), Cousins (1995) that if, for this second case, the experimenter takes the measured efficiency as exact and calculates in the standard way the "90%" C.L., then the region obtained actually corresponds to a confidence level of more than 90%. Thus, the confidence region if the efficiency is not exactly known is more restrictive than if the efficiency were exactly known. One seems to do better by knowing less!

To see how this happens, suppose the measured efficiency is ϵ_m with standard error σ, the true efficiency is ϵ_t, and that 0 events are measured. Set the upper limit for λ, the mean number of events, to be $\lambda_{U.L.}\epsilon_m \leq 2.3$. Suppose, further, that the true mean number of events is $\lambda_t = 2.3/(\epsilon_t - \delta)$, where $\delta << \sigma$.

Since δ is small, approximately 90% of the time one or more events would be seen. Consider the 10% of the time when no events are obtained. If $\epsilon_m = \epsilon_t$, then the upper limit of $2.3/\epsilon_m$ does not include λ_t, and 10% of the time an incorrect answer is obtained, as expected.

However, there is a second chance here. ϵ_m is not, in general, equal to ϵ_t. Since $\delta << \sigma$, in approximately half of the instances in which there are no events, ϵ_m will be such that $\lambda_t/\epsilon_m < 2.3$ and λ_t will be in the limit quoted. This corresponds then to a $90 + 5 = 95\%$ C.L.

This 95% CL result holds only in a narrow band of λ/ϵ_m near 2.3. The overall confidence limit must be set at the highest λ for which the appropriate percentage holds. As δ increases to and beyond σ, the effect of the second chance decreases. However, at the point at which it is negligible, one is at a somewhat higher λ/ϵ_t than 2.3, the probability of having 0 events is less than 10%, and the confidence level one can quote is higher than 90%.

This counterintuitive result occurs because of the necessary "overcoverage" for discrete distributions, that is, the actual coverage is sometimes higher than the stated coverage. If $\lambda_t < 2.3$, the upper limit will always include it (100%). If λ_t is barely above 2.3, then, in the absence of efficiency uncertainty, the upper limit will include it 90% of the time. There is a step function at 2.3. This step function becomes rounded by the uncertainty in the efficiency and at exactly 2.3, instead of a 100–90% discontinuity, the average value of 95% obtains. The average coverage, in fact, does decrease with increasing uncertainty in efficiency even though we can quote a higher minimum coverage. Although exact coverage is the ideal, overcoverage is safe, in the sense that one knows that the coverage is at least as high as the stated level.

14.5   Improving on the Symmetric Tails Confidence Limits

There has been considerable interest in finding a better procedure than just having symmetric tails for confidence intervals. Feldman and Cousins (1998) made major advances toward solving two long-standing problems concerning the use of confidence levels for estimating a parameter from data. The first of these problems is eliminating the bias that occurs when one decides between using a confidence interval or a confidence bound, after examining the data. The second is finding a confidence interval when the experimental result produces estimators that are close to or past known bounds for the parameters of interest. Feldman and Cousins' method is called the *unified approach*.

They looked at two problems. The first problem looked for a signal μ in a measurement with a Poisson distribution with expected background b, if the experiment has a signal n. They investigate the likelihood ratio $R = P(\mu|n)/P(\mu_{\text{best}}|n)$ where "best" means the μ giving the highest probability for that value of n. Next, they add values of n to the acceptance region in order of decreasing R until the sum of $P(n|\mu)$ meets or exceeds the desired confidence level. The construction proceeds by finding the acceptance region for a series of closely spaced n's. This method automatically

shifts from one-sided to two-sided limits appropriately. It yields a compact acceptance region. There is a problem if the $n < b$. They suggest looking at the sensitivity
of the result which is taken to be the average limit that can be set if there is no signal
to see how far their result is from the average.

The second problem they consider is a continuous measurement,

$$P(x|\theta) = (1/\sqrt{2\pi})e^{-(x-\theta)^2/2},$$

where θ is known to be non-negative. The probability:

$$P(x|\theta_{\text{best}}) = \frac{1}{\sqrt{2\pi}} \text{ if } x \geq 0 \text{ and } \frac{e^{-(x^2/2)}}{\sqrt{2\pi}} \text{ if } x < 0.$$

The appropriate ratio R is then

$$R(x) = \frac{P(x|\theta)}{P(x|\theta_{\text{best}})} = e^{-(x-\theta)^2/2}, x \geq 0, \text{ and } e^{(x\theta-\theta^2)/2} \text{ if } x < 0. \tag{14.10}$$

In practice this means that for a given value of θ one finds the interval $[x_1, x_2]$ such
that $R_1 = R_2$ and $\int_{x_1}^{x_2} P(x|\theta)dx = \alpha$, where α is the desired confidence level.

This unified approach does not make quite enough allowance for the known
bounds. For the Poisson problem, this was seen dramatically in an actual experiment (Boddman et al. 1998). The experiment was attempting to set limits on an
anomalous neutrino signal claimed by Athanassopoulos et al. (1996). The expected
value of the background b was $b = 2.88 \pm 0.13$ events. The experiment saw zero
events. Following the unified approach, a 90% upper limit on the neutrino signal
was set at 1.08 events, which almost excluded the results of the previous experiment.
However, the calculated sensitivity was 4.42, which indicated a possible problem. In
fact, if zero events were seen that means there were zero background events and zero
signal events. All that can be concluded from zero signal events is a 90% confidence
level of 2.3 events. The unified approach looked at the total universe of all possible
results. However, given zero events, one can look at the much smaller universe where
zero background events are obtained. Had n events been obtained, the background
is known to be at most n for this run. This suggests that the use of the conditional
probability $P(\mu|b \leq n)$ be used Roe and Woodroofe (1999). See Fig. 14.8. This mod-
ification has been particularly important for experiments searching for rare examples
of new phenomena.

The continuous problem treated by the unified method can have problems if
the best fit seems to be close to or in the unphysical negative region. A Bayesian
approach (Roe and Woodroofe 2000), gave reasonable results for both the Poisson
and the continuous problems. We start from a Bayesian approach, but will examine the results from a frequentist point of view. We assume a uniform (improper)
prior probability of 1 for $0 \leq \theta \leq \infty$. Note that the restriction to $\theta \geq 0$ makes these
examples different from those considered in Sect. 14.2.

Poisson Confidence Region (b=3.)

Fig. 14.8 The C.L. belt using the unified method (dashed line) and the Conditional probability (old RW–solid line) for the Poisson distribution with $b = 3$

Consider the continuous example. Here we measure $x = \theta + \Delta$, where $\theta \geq 0$, and the density function for x is

$$f(x|\theta) \equiv \phi(x - \theta) = \frac{1}{\sqrt{2\pi}}e^{-(x-\theta)^2/2}. \tag{14.11}$$

For the uniform prior probability, the marginal density of x becomes

$$f(x) = \int_0^\infty \phi(x - \theta)p(\theta)d\theta = \Phi(x),$$

where

$$\Phi(x) \equiv \int_{-\infty}^x \phi(y)dy,$$

the standard normal distribution function. Recall that Bayes theorem is

$$f(x|\theta) \times f(\theta) = f(\theta|x) \times f(x)$$

and that $f(\theta) = 1/\sqrt{2\pi}$ by hypothesis. The conditional density of θ given x is

$$f(\theta|x) = \frac{\phi(x - \theta)}{\Phi(x)}. \tag{14.12}$$

Note that, if we measure a value x, we know that the error Δ cannot be greater than x. $\Phi(x)$ is just the probability that $\Delta \leq x$. $f(\theta|x)$ is proper; the improper (infinite) prior probability has canceled out. We wish to find an upper limit u and a lower limit ℓ for which

$$P\{\ell \leq \theta \leq u|x\} = 1 - \epsilon. \tag{14.13}$$

It is desirable to minimize the interval $[\ell, u]$. We do that by picking the largest probability densities to be within the interval

$$[\ell, u] \doteq \{\theta : f(\theta|x) \geq c\},$$

where $0 \leq c \leq 1/\sqrt{2\pi}$ and c is chosen to give the correct interval. In Bayes theory, this interval is known as the credible interval. $f(\theta|x) \geq c$ if and only if $|\theta - x| \leq d$, where

$$d = \sqrt{-2 \ln c - \ln(2\pi) - 2 \ln[\Phi(x)]}.$$

There are two cases to be considered.

(1) $d \leq x$. Then the condition Eq. 14.11, becomes

$$1 - \epsilon = \int_{x-d}^{x+d} f(\theta|x)d\theta = \int_{-d}^{d} \frac{\phi(\omega)}{\Phi(x)} d\omega = \frac{2\Phi(d) - 1}{\Phi(x)};$$

$$d = \Phi^{-1}[\frac{1}{2} + \frac{1}{2}(1 - \epsilon)\Phi(x)]. \tag{14.14}$$

(2) $x < d$. Equation 14.11 becomes

$$\epsilon = \int_{x+d}^{\infty} f(\theta|x)d\theta = \int_{d}^{\infty} \frac{\phi(\omega)}{\Phi(x)} d\omega = \frac{1 - \Phi(d)}{\Phi(x)};$$

$$d = \Phi^{-1}[1 - \epsilon\Phi(x)]. \tag{14.15}$$

If x_0 is the point where these two curves meet, then, after a little algebra,

$$x_0 = \Phi^{-1}\left(\frac{1}{1 + \epsilon}\right). \tag{14.16}$$

Hence

$$[\ell, u] = [\max(x - d, 0), x + d], \quad \text{where}$$

$$d = \Phi^{-1}[1 - \epsilon\Phi(x)] \quad \text{if} \quad -\infty < x \le x_0,$$
$$= \Phi^{-1}[1/2 + (1/2)(1 - \epsilon)\Phi(x)] \quad \text{if} \quad x_0 < x < \infty.$$

Consider the coverages. From a Bayesian point of view, this is an exact credible region with minimized intervals. Next, examine the frequentist point of view. Fix θ and find the points at this fixed θ where x meets the upper and lower Bayesian limits. Let x_ℓ be that x which meets the upper limit, $\theta = u(x_\ell) = x_\ell + d(x_\ell)$, and x_u be that x which meets the lower limit, $\theta = \ell(x_u) = x_u - \min[d(x_u), x_u]$.

It can be shown that for the error $\theta - x$, these intervals correspond to exact conditional coverage at the $1 - \epsilon$ level. If x' is an independent sample picked from the conditional sample space, then $Prob[x_\ell - x \le \theta - x' \le x_u - x | \Delta \le x] = 1 - \epsilon$. The conventional coverage (credible coverage) is then necessarily not exact. A lower limit on the coverage can be shown to be

$$\Phi(x_u - \theta) - \Phi(x_\ell - \theta) \ge \frac{1 - \epsilon}{1 + \epsilon},$$

but this is a very conservative limit. For $\epsilon = 0.1$, this lower limit is 0.8182, while a numerical calculation finds a minimum coverage of about 0.86.

The conventional coverage probability can be improved for a very small increase in limits. We will consider an alternate Bayesian upper limit on θ obtained as the one-sided limit for a higher confidence level, $\epsilon' = \epsilon/2$. Let

$$u'(x) = \max\left[u(x), x + \Phi^{-1}\left(1 - \frac{1}{2}\epsilon\right)\right].$$

Let x'_ℓ be the x corresponding to $u'(x) = \theta$. The coverage can then be shown to be given by

$$\Phi(x_u - \theta) - \Phi(x'_\ell - \theta) = \frac{1}{2} + \frac{1}{2}(1 - \epsilon)\Phi(x_u) - \min\left[\frac{1}{2}, \Phi(x'_\ell)\right]\epsilon.$$

The under-coverage is very small. For a 90% credible region, the conventional coverage is at least 0.900 everywhere to three significant figures. The confidence bands are shown in Fig. 14.9.

Next, consider the discrete example. The Poisson distribution is

$$p(n|\theta) = \frac{e^{-\lambda}\lambda^n}{n!}. \tag{14.17}$$

For our present problem $\lambda = \theta + b$, where $b \ge 0$ is the fixed "background" mean. We then have

Confidence Belt For Continuous Example

Fig. 14.9 The 90% C.L. belt using the Bayesian procedure and the improved region (dashed) for the continuous example

$$p(n) = \int_0^\infty \frac{1}{n!}(\theta+b)^n e^{-(\theta+b)} p(\theta+b)d\theta = \frac{e^{-b}}{n!}\int_0^\infty \sum_{k=0}^n \theta^k b^{n-k}\frac{n!}{k!(n-k)!}e^{-\theta}d\theta$$

$$= \sum_{k=0}^n b^{n-k}\frac{e^{-b}}{k!(n-k)!}\int_0^\infty \theta^k e^{-\theta}d\theta = \sum_{k=0}^n \frac{b^{n-k}e^{-b}}{(n-k)!}$$

$$= \sum_{\ell=0}^n \frac{b^\ell e^{-b}}{\ell!} = P(\leq n \text{ counts}|b) \equiv P_b(n),$$

$$(14.18)$$

where we noted that $\int_0^\infty \theta^k e^{-\theta}d\theta = k!$. We are using p for differential probability and P for integral probability. Use Bayes theorem $p(n|\theta+b) \times p(\theta+b) = p(n, \theta+b) = p(\theta+b|n) \times p(n)$. Then, recalling that $p(\theta) = 1$ for $\theta \geq 0$ and $p(n) = P_b(n)$, we have

$$p(\theta+b|n) = \frac{p(n|\theta+b)}{P_b(n)}.$$

$$(14.19)$$

Set the upper limit for $\theta = u$, and the lower limit $= \ell$. We wish to find the credible interval $P\{\ell \le \theta \le u\} = 1 - \epsilon$, where the probability that θ is not in the interval is ϵ.

We wish to minimize $u - \ell$, in order to obtain the shortest region possible. This is accomplished by including the highest probabilities possible within the region, i.e., the interval $[\ell, u] = \{\theta : p(\theta + b|n) \ge c\}$. $p(\theta + b|n) \ge c$ if

$$(\theta + b)^n e^{-(\theta+b)} \ge P_b(n)n!c. \tag{14.20}$$

The maximum probability for physical θ, i.e., $\theta \ge 0$, occurs at

$$\frac{d}{d\theta}(\theta + b)^n e^{-(\theta+b)} = 0 = e^{-(\theta+b)}[n(\theta + b)^{n-1} - (\theta + b)^n]. \tag{14.21}$$

Thus

$$\theta_{max} = \max(n - b, 0). \tag{14.22}$$

The maximum value of c is then

$$
\begin{aligned}
c_{max} &= \frac{e^{-n}n^n}{n!P_b(n)}, \quad \text{if } n - b \ge 0 \\
&= \frac{e^{-b}b^n}{n!P_b(n)} \quad \text{otherwise.}
\end{aligned}
\tag{14.23}
$$

Let c_{min} be the c corresponding to $\theta = 0$.

$$c_{min} = \frac{e^{-b}b^n}{n!P_b(n)}. \tag{14.24}$$

The probability of the credible region is $1 - \epsilon$. For a 90% credible region, $\epsilon = 0.1$. If there is an upper and a lower limit, then we can calculate the credible coverage.

$$
\begin{aligned}
1 - \epsilon &= \int_\ell^u \frac{p(n|\theta, b)}{P_b(n)} d\theta \\
&= \frac{e^{-b}}{P_b(n)n!} \int_\ell^u (\theta + b)^n e^{-\theta} d\theta \\
&= \frac{e^{-b}}{P_b(n)n!} \sum_{k=0}^n \frac{b^k n!}{k!(n-k)!} \int_\ell^u \theta^{(n-k)} e^{-\theta} d\theta.
\end{aligned}
\tag{14.25}
$$

Now $\int_\ell^u = \int_\ell^\infty - \int_u^\infty$. Consider $\int_a^\infty \theta^{n-k} e^{-\theta} d\theta$ and set $\theta' = \theta - a$. Then $\int_a^\infty \theta^{n-k} e^{-\theta} d\theta = e^{-a} \int_0^\infty (\theta' + a)^{n-k} e^{-\theta'} d\theta'$. Expand $(\theta' + a)^{n-k}$.

$$\int_a^\infty \theta^{n-k} e^{-\theta} d\theta = \sum_{j=0}^{n-k} \frac{a^{n-k-j}(n-k)! e^{-a}}{(n-k-j)! j!} \int_0^\infty (\theta')^j e^{-\theta'} d\theta'. \tag{14.26}$$

Note again that $\int_0^\infty (\theta')^j e^{-\theta'} d\theta' = j!$. Then

$$\int_\ell^u \theta^{n-k} e^{-\theta} d\theta = \sum_{j=0}^{n-k} \frac{(\ell^{n-k-j} e^{-\ell} - u^{n-k-j} e^{-u})(n-k)!}{(n-k-j)!}. \tag{14.27}$$

$$
\begin{aligned}
1 - \epsilon &= \frac{e^{-b}}{P_b(n)} \sum_{k=0}^n \frac{b^k}{k!} \sum_{j=0}^{n-k} \frac{\ell^{n-k-j} e^{-\ell} - u^{n-k-j} e^{-u}}{(n-k-j)!} \\
&= \frac{1}{P_b(n)} \sum_{k=0}^n p(k|b) \sum_{j=0}^{n-k} [p(n-k-j|\ell) - p(n-k-j|u)] \tag{14.28} \\
&= \frac{1}{P_b(n)} [P_{\ell+b}(n) - P_{u+b}(n)].
\end{aligned}
$$

This is the credible region. Calculation is done numerically, by iteration.

The resulting region, by construction, minimizes the intervals in the Bayesian sense. Consider next, the frequentist interpretation. The conditional coverage can be shown to be $\approx 1 - \epsilon$, except for discreteness. The conventional coverage is not exact. For $b = 3$ and $1 - \epsilon = 0.9$, for example, the conventional coverage varies from about 86–96.6%.

The conventional coverage can be improved by a small ad hoc modification similar to that used for the continuous case. Consider an alternate Bayesian upper limit u' for θ defined as the one-sided limit for a confidence level C.L.$'$ $> 1 - \epsilon$. Take the Bayesian upper limit as the maximum of u and u'. For $b = 3$, and a 90% C.L., a C.L.$'$ of 92% yields a conventional C.L. $>$88%. The value of the limit for $n = 0$ then increases from 2.3 to 2.53 and is independent of b for fixed C.L.$'$. However, it is probably useful to tune C.L.$'$ depending on b.

Although this Bayesian result is not as accurate as the conditional confidence method, it becomes natural to use it if there is some partial knowledge of the background-signal separation, which is the subject of the next section.

14.6 Improvement over Symmetric Tails Confidence Limits for Events With Partial Background-Signal Separation

In the last section, we discussed a method of choosing a confidence belt for events from a Poisson distribution in which the mean was the sum of the desired, unknown signal θ and the known background b. The method we discussed had several advantages over just choosing symmetric tails. We now wish to extend this method to processes in which we have some information concerning whether a given event is a signal or a background event.

Suppose on each event one measures a statistic x (or a vector of statistics) for which the density function is $g(x)$ for signal events and $h(x)$ for background events. Suppose, further, that the number of observed events is n, and that these are distributed in a Poisson distribution with mean $b + \theta$. θ is the expected number of signal events. We assume that b, the expected number of background events, is known. Let $j_k = 1, 0$ depending on whether the kth event is a signal or a background event. Let x_k be the parameter measured for the kth event. Initially, suppose that we know which individual events are signal and which are background. Then

$$\frac{d^n p}{dx_1 \cdots dx_n}\{n, j_k, x_k, 1 \le k \le n | \theta\}$$
$$= \frac{\theta^m b^{n-m}}{n!} e^{-(b+\theta)} \prod_{k=1}^{n} g^{j_k}(x_k) h^{1-j_k}(x_k), \tag{14.29}$$

where $m = \sum_{k=1}^{n} j_k$ is the number of signal events. Note that we have divided the product of the Poisson probabilities by the binomial coefficient. The Poisson probabilities give the probability that there are m signal and $n - m$ background events in any order and the probability given in the preceding equation is for one particular order. Next, we sum over the various arrangements which give m signal events

$$\frac{d^n p}{dx_1 \cdots dx_n}\{n, m, x_k, 1 \le k \le n | \theta\}$$
$$= \sum_{j_1+\cdots+j_n=m} \frac{d^n p}{dx_1 \cdots dx_n}\{n, j_k, x_k, 1 \le k \le n | \theta\} \tag{14.30}$$
$$= \frac{\theta^m b^{n-m}}{m!(n-m)!} e^{-(b+\theta)} C_{n,m},$$

where

$$C_{n,m} = \left[1 / \binom{n}{m}\right] \sum_{j_1+j_2+\cdots+j_n=m} \prod_{k=1}^{n} g^{j_k}(x_k) h^{1-j_k}(x_k). \tag{14.31}$$

Note that if $g = h = 1$, then $C_{n,m} = 1$ and we recover the results obtained if no partial information is known. Next, assume the actual situation in which n and the x's are known, but m is not known

$$f(n, x|\theta) \equiv \frac{d^n p}{dx_1 \cdots dx_n}\{n, x_k, k = 1, n|\theta\}$$

$$= \frac{(\theta + b)^n}{n!} e^{-(\theta+b)} \prod_{i=1}^{n} \frac{\theta g(x_i) + bh(x_i)}{\theta + b}. \tag{14.32}$$

As in the last section, we will use Bayes theorem, but then interpret the results in terms of frequentist limits. Let the prior probability be the improper prior $p(\theta) = 1$ for $\theta \geq 0$. Then

$$f(n, x) \equiv \int_0^\infty f(n, x|\theta) d\theta = \int_0^\infty \frac{1}{n!} e^{-(\theta+b)} \prod_{i=1}^{n} [\theta g(x_i) + bh(x_i)] d\theta. \tag{14.33}$$

$$f(n, x) = \frac{e^{-b}}{n!} \sum_{m=0}^{n} b^{n-m} \int_0^\infty \theta^m e^{-\theta} C_{n,m} \binom{n}{m} d\theta.$$

Recall that $\int_0^\infty \theta^m e^{-\theta} d\theta = m!$ and $\binom{n}{m} = n!/(m!(n-m)!$

$$f(n, x) = \sum_{m=0}^{n} \frac{e^{-b} b^{n-m}}{(n-m)!} C_{n,m}. \tag{14.34}$$

$$g(\theta|n, x) \equiv p\{\theta|n, x_k, k = 1, n\}$$

$$= \frac{(1/n!)e^{-(\theta+b)} \prod_{i=1}^{n}[\theta g(x_i) + bh(x_i)]}{\sum_{m=0}^{n} e^{-b} \frac{b^{n-m}}{(n-m)!} C_{n,m}}, \tag{14.35}$$

$$g(\theta|n, x) = \frac{e^{-\theta} \prod_{i=1}^{n}[\theta g(x_i) + bh(x_i)]}{n! \sum_{m=0}^{n} \frac{b^{n-m}}{(n-m)!} C_{n,m}}. \tag{14.36}$$

We want to find upper and lower limits u, ℓ for θ such that $Prob\{\ell \leq \theta \leq u|n, x\} = 1 - \epsilon$ and to minimize that interval. This means we want $[\ell, u] = \{\theta : g(\theta|n, x) \geq c\}$. Note that the denominator

$$D = n! \sum_{m=0}^{n} \frac{b^{n-m}}{(n-m)!} C_{n,m} \tag{14.37}$$

is independent of θ. First we find the $\theta = \theta_{max}$, for which g is maximum

$$0 = \frac{d}{d\theta} \frac{e^{-\theta}}{D} \prod_{i=1}^{n} [\theta g(x_i) + bh(x_i)]$$

$$= \frac{e^{-\theta}}{D} \left\{ -\prod_{i=1}^{n} [\theta g(x_i) + bh(x_i)] + \sum_{j=1}^{n} g(x_j) \prod_{i=1}^{n} [\theta g(x_i) + bh(x_i)]^{1-\delta_{ij}} \right\}$$

$$= \frac{e^{-\theta}}{D} \prod_{i=1}^{n} \sum_{j=1}^{n} \left\{ \left[\frac{g(x_j)}{\theta g(x_j) + bh(x_j)} - \frac{1}{n} \right] [\theta g(x_i) + bh(x_i)] \right\}$$

$$= \frac{e^{-\theta}}{D} \left\{ \prod_{i=1}^{n} [\theta g(x_i) + bh(x_i)] \right\} \left\{ \sum_{j=1}^{n} \left[\frac{g(x_j)}{\theta g(x_j) + bh(x_j)} - \frac{1}{n} \right] \right\}.$$

$$(14.38)$$

Thus

$$\sum_{j=1}^{n} \left[\frac{g(x_j)}{\theta_{max} g(x_j) + bh(x_j)} \right] = 1 \qquad (14.39)$$

defines θ_{max} if θ_{max} is > 0. Otherwise, $\theta_{max} = 0$. Next, integrate $g(\theta|n, x)$. Let

$$G(a) = \int_0^a g(\theta|n, x)d\theta = 1 - \int_a^\infty g(\theta|n, x)d\theta. \qquad (14.40)$$

$$G(a) \equiv \int_0^a g(\theta|n, x)d\theta = \frac{1}{D} \int_0^a e^{-\theta} \prod_{i=1}^{n} [\theta g(x_i) + bh(x_i)]d\theta$$

$$(14.41)$$

$$= \frac{1}{D} \sum_{m=0}^{n} b^{n-m} C_{n,m} \binom{n}{m} m! \int_0^a \frac{\theta^m}{m!} e^{-\theta} d\theta.$$

$$1 - G(a) = \frac{1}{D} \sum_{m=0}^{n} b^{n-m} C_{n,m} \binom{n}{m} \int_a^\infty \theta^m e^{-\theta} d\theta.$$

Use the result given in Eq. 14.23, to evaluate this integral. We then obtain

$$1 - G(a) = \frac{1}{D} \sum_{m=0}^{n} b^{n-m} C_{n,m} \frac{n!}{m!(n-m)!} m! e^{-a} \sum_{l=0}^{m} \frac{a^l}{l!}$$

$$(14.42)$$

$$= \frac{e^{-a} \sum_{i=0}^{n} \frac{b^{n-i} C_{n,i}}{(n-i)!} \sum_{l=0}^{i} \frac{a^l}{l!}}{\sum_{m=0}^{n} \frac{b^{n-m}}{(n-m)!} C_{n,m}}.$$

We can use either Eq. 14.41, recognizing that the integral is an incomplete gamma function, or use Eq. 14.41, to find $G(a)$. The limits can then be found by iterations as follows:

(1) $\theta_{max} = 0$. By iteration find a such that $G(a) = 1 - \epsilon$. This is the upper limit. The lower limit is 0.
(2) $\theta_{max} > 0$. Find $g(\theta = 0|n, x)$. By iteration find
$g([\{\theta = a\} > \theta_{max}]|n, x) = g(0|n, x)$. Calculate $G(a)$.

- If $G(a) \leq 1 - \epsilon$, then the lower limit is 0 and proceed as in 1.).
- If $G(a) > 1 - \epsilon$, then there is a two-sided limit. Iterate trying larger lower limits, finding an upper limit with the same probability to zero in on Probability $= G(u) - G(\ell) = 1 - \epsilon$.

14.7 When Is a Signal Significant?

We now turn to a subject that is hard to quantify but is very important in practice. I will use an example from particle physics, but the problem is universal. Suppose I have examined a set of particle interactions and am searching for new particles (i.e., resonances). I look at the effective mass spectrum of $\pi^+\pi^+\pi^-$. Suppose I have a histogram with 100 bins each of which is 25 MeV wide. One bin has 40 events, while the nearby bins average 20 events. Is this a signal?

This is a 4.5 standard deviation peak. Using the Poisson distribution, the probability of obtaining ≥ 40 events when 20 are expected is about 5.5×10^{-5}. It would seem that this is very significant. But is it? Suppose I do not know in advance which bin might have the resonance. Then the signal could have been in any of the 100 bins in the histogram. The probability of this size signal in some bin is $100 \times 5.5 \times 10^{-5} = 5.5 \times 10^{-3}$. Suppose I had also looked at a number of other mass spectra from these reactions searching for a resonance and found only this one bump. For example, suppose I had plotted 30 different histograms to look for resonances ($\pi^+\pi^+\pi^-$, $\pi^+\pi^-\pi^0$, $K^+\pi^+\pi^-$, ...). The probability of this size bump in any of those is $5.5 \times 10^{-3} \times 30 = 0.16$, which is not unusual at all. Physicists have often been fooled by not taking these sorts of considerations into account, and physics literature is filled with four to five standard deviation bumps that proved to be false.

A further problem with the above example is that if the nearby bins are averaged including the high bin, the background would be higher than 20 events. By omitting this bin, we bias the data if there is no signal. To test the hypothesis of "no real bump", we must include the suspect bin in calculating the size of background expected.

Another related very common trap occurs as follows. I think I see a bump somewhere. Are there some data cuts that will make it stand out better? Perhaps it is produced mainly at low t. Therefore, I look at events with low t and find a better signal. By continuing in this way, I soon have a fine signal. The problem is that the cuts were chosen to enhance the signal. If the low t cut had produced a poorer signal, I would have concluded that the resonance was produced at high t and selected high t events. This is biased.

What can be done? It is often not clear how to calculate the effects of bins and bias. In the first example, the effective number of bins, etc., may be debatable. Of course, if I had stated in advance of the experiment that I expected a peak at 2.1 GeV for low t events, then the procedure is fine and the probability correct.

In one case some years ago a friend of mine (G. Lynch) was faced with a situation of whether the data of a histogram had a peak (over several bins). He made up a set of 100 Monte Carlo histograms with no intrinsic peak, mixed the real histogram with it, and gave it to members of his group to decide which histograms had the most prominent peak. The real data came in third and no resonance was claimed.

In another case, in an international collaboration, we had a group of enthusiasts who had made various cuts and produced a very unexpected $\mu\pi$ resonance in neutrino interactions using the first half of our data. We had lots of arguments about whether it was publishable. Fortunately, we were about to analyze the second half of our data. We froze the cuts from the first half and asked whether they produced the same peak in the second half. This is a fair procedure and is useful if you have enough data.

Play with one-half of the data and then if an effect exists, check it in the second half. It is still necessary to derate the probability by the number of times the second half got checked, but it is a considerable help. In our particular case, the signal vanished, but with similar but not quite the same cuts, a new signal could be found in μk.

A variant of this occurs when looking for a new effect in data, which is seen by comparison with a Monte Carlo estimate. Cuts should be frozen by using just the Monte Carlo, then comparing with data.

To summarize, if you search for a bump in any of a number of histograms, remember to derate the result by the number of bins in which a peak might have appeared in all the histograms you examined. Remember to calculate the background height including the peak bin to test the hypothesis of "no real bump". If you have made cuts to enhance the bump, you must examine an independent set of data with the same cuts.

The problem of bias comes in many forms, of which we have examined only a few instructive examples here. You should keep this sort of problem in mind when examining data. How should you proceed generally? My suggestion is that privately and within your group, be an enthusiast. Try all the cuts. It would be exceedingly unfortunate if you let these considerations stop you from finding real effects. However, after you have gone through these procedures and produced your signal, you should look at the result with a skeptical, realistic eye. It is often very difficult to do this in a competitive environment with the excitement of a possible discovery. It is, however, very important and is one of the hallmarks of a mature physicist.

A number of specific "look elsewhere" tests are discussed in Chap. 19.

In this chapter, we have examined the problems of inverse probability, estimating parameters from data. We have examined uses (and occasional misuses) of Bayes theorem and have introduced confidence limits as a means of avoiding many of the subtle problems in this subject. It is extremely important to frame the question you ask carefully—and to understand the answer in the light of the precise question asked.

14.8 Worked Problems

WP14.1(a) Complete the example of beam particle velocity, i.e., show that the equation leads to a normal distribution and find the mean and variance.

Answer:

We have

$$f(v|v_1)\,dv = \frac{\frac{1}{\sqrt{2\pi\sigma_1^2}}\exp\left(-\frac{(v-v_1)^2}{2\sigma_1^2}\right)\frac{1}{\sqrt{2\pi\sigma_0^2}}\exp\left(-\frac{(v-v_0)^2}{2\sigma_0^2}\right)dv}{\int_{-\infty}^{\infty}\frac{1}{\sqrt{2\pi\sigma_1^2}}\exp\left(-\frac{(v-v_1)^2}{2\sigma_1^2}\right)\frac{1}{\sqrt{2\pi\sigma_0^2}}\exp\left(-\frac{(v-v_0)^2}{2\sigma_0^2}\right)dv}.$$

Consider the argument of the exponential

$$= -\frac{(v-v_1)^2}{2\sigma_1^2} - \frac{(v-v_0)^2}{2\sigma_0^2}$$

$$\text{Argument} = -\left(v^2\left(\frac{1}{2\sigma_1^2}+\frac{1}{2\sigma_0^2}\right) - v\left(\frac{v_1}{\sigma_1^2}+\frac{v_0}{\sigma_0^2}\right) + \frac{v_1^2}{2\sigma_1^2}+\frac{v_0^2}{2\sigma_0^2}\right).$$

$$\frac{1}{\sigma^2} = \frac{1}{\sigma_1^2} + \frac{1}{\sigma_0^2}.$$

$$\text{Let Argument} = -\frac{1}{2\sigma^2}\left(v^2 - 2\sigma^2\left(\frac{v_1}{\sigma_1^2}+\frac{v_0}{\sigma_0^2}\right)v + \sigma^2\left(\frac{v_1^2}{\sigma_1^2}+\frac{v_0^2}{\sigma_0^2}\right)\right)$$

$$\text{The Argument} = -\frac{1}{2\sigma^2}\left(v - \left(\frac{v_1}{\sigma_1^2}+\frac{v_0}{\sigma_0^2}\right)\sigma^2\right)^2 + C,$$

$$(14.43)$$

where C is independent of v, and therefore, only effects the normalization, which is guaranteed to be correct by the initial expression we wrote down for $f(v|v_1)$. Hence, we see that $f(v|v_1)$ is indeed a normal frequency function with

$$v_m \equiv \text{mean} = \left(\frac{v_1}{\sigma_1^2}+\frac{v_0}{\sigma_0^2}\right)\sigma^2 = \frac{v_1/\sigma_1^2 + v_0/\sigma_0^2}{1/\sigma_1^2 + 1/\sigma_0^2},$$

and with variance $= \sigma^2$. Thus

$$f(v|v_1)\,dv = \frac{1}{\sqrt{2\pi\sigma^2}}\exp\left(-\frac{1}{2\sigma^2}(v-v_m)^2\right)dv.$$

WP14.1(b) Show the mean and variance are the same as that obtained by looking at the problem as having two measurements of v and obtaining the best weighted mean.

Answer:

By comparing the results of part *a* with the results of Worked Problem 3.2, we see that the form of v_m and σ^2 above is identical with the form we found for the weighted mean of two measurements.

WP14.1(c) Find the mean and variance for the best estimate if

$$v_0 = 50 \times 10^6 \text{ m/s}, \qquad \sigma_0^2 = 1 \times 10^{12} \text{ (m/s)}^2,$$
$$v_1 = 55 \times 10^6 \text{ m/s}, \qquad \sigma_1^2 = 2 \times 10^{12} \text{ (m/s)}^2. \tag{14.44}$$

Answer:

$$v_0 = 50 \times 10^6 \text{ m/s}, \ \sigma_0^2 = 1 \times 10^{12} \text{ (m/s)}^2.$$
$$\text{If } v_1 = 55 \times 10^6 \text{ m/s}, \ \sigma_1^2 = 2 \times 10^{12} \text{ (m/s)}^2, \tag{14.45}$$

then

$$\sigma^2 = \frac{1}{\frac{1}{1\times10^{12}} + \frac{1}{2\times10^{12}}} = \frac{2}{3} \times 10^{12} \left(\frac{m}{s}\right)^2,$$
$$v_m = \left(\frac{50 \times 10^6}{1 \times 10^{12}} + \frac{55 \times 10^6}{2 \times 10^{12}}\right) \frac{2}{3} \times 10^{12} = 51.67 \times 10^6 \text{ m/s}. \tag{14.46}$$

WP14.2 A new charmed particle is seen in an emulsion stack-electronics detector array exposed to a neutrino beam. By measuring the energy of the secondaries, it is determined that for this particular event the charmed particle lived for 3×10^{-13} sec in its rest system before decay.
What are the symmetric 90% confidence level limits on the lifetime of this kind of particle?

Answer:

For a given τ

$$f(t) \, dt = \frac{1}{\tau} e^{-t/\tau} \, dt,$$
$$\int_{t_1}^{t_2} f(t) \, dt = e^{-t_1/\tau} - e^{-t_2/\tau}. \tag{14.47}$$

We want to look at 5% on each end, i.e.,
Case 1.   Let $t_2 \to \infty$. Then $e^{-t_1/\tau} = 0.95$, i.e., there is only a 5% chance that t_1 is less than this value for fixed τ.

$$-t_1/\tau = \log .95 \Longrightarrow \tau = 5.85 \times 10^{-12} \text{ sec for the observed } t_1 \text{ value.}$$

Case 2. Let $t_1 \to 0$. Then $1 - e^{-t_2/\tau} = 0.95$, i.e., there is only a 5% chance that t_2 is greater than this value for fixed τ.

$$e^{-t_2/\tau} = .05 \Longrightarrow \tau = 1 \times 10^{-13} \text{ sec.}$$

What would the Bayesian approach give?

$$g(\tau) \, d\tau = \frac{\frac{1}{\tau}e^{-t/\tau} * \text{const } d\tau}{\int_0^\infty \text{const} \frac{1}{\tau}e^{-t/\tau} d\tau}. \tag{14.48}$$

$$I = \int_0^\infty \frac{1}{\tau}e^{-t/\tau} \, d\tau.$$

$$y = \frac{t}{\tau}, \ \tau = \frac{t}{y}, \ \text{or } d\tau = \frac{-t}{y^2} \, dy. \tag{14.49}$$

$$Let\, I = \frac{-t}{t} \int_\infty^0 ye^{-y} \frac{dy}{y^2} = \int_0^\infty \frac{1}{y}e^{-y} \, dy = \infty.$$

This diverges at the lower limit! Here the desperation hypothesis is not even well defined.

WP14.3 We are attempting to determine the polarization of a beam of Λ particles. In the center-of-mass system, the decay probability is proportional to $(1 + \beta \cos \phi) \, d\cos \phi$, where ϕ is the angle with the line of flight (i.e., the direction of motion of the lab as seen in the center of mass). β will provide us with a measure of polarization. Of 100 Λ observed, 68 decay forward ($\cos\phi > 0$). What are the symmetric 95% confidence level limits for β? (Use Fig. 14.6.)

Answer:

$$P_{forward}\alpha \int_0^1 (1 + \beta \cos \phi) \, d\cos \phi = \int_0^1 (1 + \beta x) \, dx = 1 + \beta/2.$$

Note that the integral over the full angular range is

$$\int_{-1}^1 (1 + \beta x) \, dx = 2.$$

The observed forward decay probability, $p = \alpha^* = 0.68$. Figure 14.6 has 0.54–0.78 as limits for this value. $(1 + \beta/2)/2 = p$ or $\beta = 4p - 2$. Thus, $0.16 \leq \beta \leq 1.12$ are the limits. However, 1.12 is unphysical. Hence, the limit is $\beta > 0.16$. We have excluded the region $-1 \leq \beta \leq 0.16$.

14.9   Exercises

14.1 Derive and plot the symmetrical confidence limit curves shown in Fig. 14.6, for the binomial distribution for the case $n = 20$. (If you use a computer, you might as well have it plot the whole family of curves.)

14.2 Popcorn is popping in a hot pan and a measurement made of how high each kernel pops. In a given batch, 25 kernels are observed to pop more than 4 in. high. If the total number of kernels is considerably larger, indicate qualitatively why this problem satisfies the conditions for a Poisson distribution. Determine the symmetric 95% confidence level limits on the mean number of kernels expected to pop above 4 in. for such a batch. (Note that rather similar considerations can be applied to the number of molecules evaporating from a surface in a given time.)

14.3 Show that if the probability of obtaining n events is given by the Poisson distribution with mean θ and the probability of θ is given by the gamma distribution with mean α/μ, and variance α/μ^2, then the overall probability of obtaining n events is given by the negative binomial distribution as discussed in Sect. 14.3, and that the parameters of the negative binomial distribution are $m = \alpha$, $r = n + \alpha$, and $p = \mu/(\mu + 1)$.

References

Athanassopoulos C et al (1996) Phys Rev C 54:2685
Boddman B et al (1998) Results from the karmen neutriino experiment. Nucl Phys B (Proc Suppl) 31:301–306
Cousins R (1995) Am J Phys 63:398
Cousins R, Highland V (1992) Nuc Inst Meth A320:331
DeLury D, Chung J (1950) Confidence limits for the hypergeometric distribution. University of Toronto Press, Toronto
Feldman G, Cousins R (1998) Unified approach to the classical statistical analysis of small signals. Phys Rev D 57:3873
Lehman E (1986) Testing statistical hypotheses, 2nd edn. Wiley, New York
Neyman J (1937) Philos Trans R Soc Lond A 236:333
Parratt L (1961) Probability and experimental errors in science; an elementary survey. Wiley, New York
Roe B, Woodroofe M (1999) Improved probability method for estimating signal in the presence of background. Phys Rev D 60(053):009
Roe B, Woodroofe M (2000) Setting confidence belts. Phys Rev, 13009

Chapter 15
Curve Fitting

Abstract In this chapter, we will look at the problem of fitting data to curves and estimating several parameters. For the often-occurring linear case, we can turn the procedure into a crank-turning procedure. If the dependence on the parameters is intrinsically non-linear, we will see that the problem is much harder, but general computer programs to find minima of multidimensional functions can be of considerable help. Regression analysis with a non-constant variance will be discussed. The problem of instabilities associated with the Gibbs phenomenon and the reason for the problem will be discussed and various methods of regularization to attack the problem will be examined. Problems of fitting data with errors in both x and y will be discussed as will the problems of non-linear parameters. Finally the questions of optimizing a data set with signal and background, and of the robustness of estimates will be examined.

15.1 The Maximum Likelihood Method for Multiparameter Problems

We will first examine the use of the maximum likelihood method for multiparameter problems. α_j^* is the jth parameter to be fit and α_{j_0} is the expectation value of that parameter. It is assumed that the correlation between two parameters, the moment matrix $\Lambda_{ij} = \mathbb{E}[(\alpha_i^* - \alpha_{i_0})(\alpha_j^* - \alpha_{j_0})]$ is known. In one dimension, $\Lambda = \sigma^2$. The analog of the basic theorem quoted in the last chapter is that the distribution of $\alpha_j^* - \alpha_{j_0}$, $j = 1, 2, \ldots, s$ approaches the s-dimensional normal distribution

$$\frac{1}{(2\pi)^{s/2}\sqrt{\Lambda}}\exp\left(-\frac{(\alpha_i^* - \alpha_{i_0})(\Lambda^{-1})_{ij}(\alpha_j^* - \alpha_{j_0})}{2}\right), \tag{15.1}$$

where Λ is the determinant of Λ_{ij}. Furthermore, for a set of n independent identical trials, it can be shown that

$$(\Lambda^{-1})_{ij} = n\mathbb{E}\left\{\frac{\partial \log f}{\partial \alpha_i}\frac{\partial \log f}{\partial \alpha_j}\right\}. \tag{15.2}$$

© Springer Nature Switzerland AG 2020
B. P. Roe, *Probability and Statistics in the Physical Sciences*,
Undergraduate Texts in Physics,
https://doi.org/10.1007/978-3-030-53694-7_15

If the trials are independent but not necessarily identical,

$$(\Lambda^{-1})_{ij} = \sum_{r=1}^{n} \mathbb{E} \left\{ \frac{\partial \log f_r}{\partial \alpha_i} \frac{\partial \log f_r}{\partial \alpha_j} \right\}. \tag{15.3}$$

Two other forms of the above relation for Λ^{-1} are very useful. Remember

$$w = \log \mathcal{L} = \sum_{r=1}^{n} \log f_r, \tag{15.4}$$

$$\frac{\partial w}{\partial \alpha_i} = \sum_{r=1}^{n} \frac{\partial \log f_r}{\partial \alpha_i}, \tag{15.5}$$

$$\frac{\partial w}{\partial \alpha_i} \frac{\partial w}{\partial \alpha_j} = \sum_{r_1=1}^{n} \sum_{r_2=1}^{n} \frac{\partial \log f_{r_1}}{\partial \alpha_i} \frac{\partial \log f_{r_2}}{\partial \alpha_j}. \tag{15.6}$$

The cross terms will have zero expectation value.

We see this as follows. For a cross term with $r_1 \neq r_2$, f_{r_1} and f_{r_2} are independent. Therefore,

$$\frac{\partial \log f_{r_1}}{\partial \alpha_i} * \frac{\partial \log f_{r_2}}{\partial \alpha_j} = \frac{\partial \log f_{r_1}}{\partial \alpha_i} * \frac{\partial \log f_{r_2}}{\partial \alpha_j}.$$

However,

$$\frac{\partial \log f_{r_1}}{\partial \alpha_i} = \int \frac{1}{f_{r_1}} \frac{\partial f_{r_1}}{\partial \alpha_i} f_{r_1} \, dx = \frac{\partial}{\partial \alpha} \int f_{r_1} \, dx$$

$$= \frac{\partial}{\partial \alpha}(1) = 0.$$

$$\mathbb{E} \left\{ \frac{\partial w}{\partial \alpha_i} \frac{\partial w}{\partial \alpha_j} \right\} = \sum_{r=1}^{n} \mathbb{E} \left\{ \frac{\partial \log f_r}{\partial \alpha_i} \frac{\partial \log f_r}{\partial \alpha_j} \right\}. \tag{15.7}$$

Thus,

$$(\Lambda^{-1})_{ij} = \mathbb{E} \left\{ \frac{\partial w}{\partial \alpha_i} \frac{\partial w}{\partial \alpha_j} \right\}. \tag{15.8}$$

Next we consider

$$\frac{-\partial^2 w}{\partial \alpha_i \partial \alpha_j} = -\sum_{r=1}^{n} \frac{\partial^2 \log f_r}{\partial \alpha_i \partial \alpha_j}. \tag{15.9}$$

For simplicity of notation, assume all of the r trials are identical.

$$\mathbb{E}\left\{-\frac{\partial^2 w}{\partial \alpha_i \, \partial \alpha_j}\right\} = n\mathbb{E}\left\{-\frac{\partial^2 \log f}{\partial \alpha_i \, \partial \alpha_j}\right\},$$

$$\frac{\partial^2 \log f}{\partial \alpha_i \, \partial \alpha_j} = \frac{\partial}{\partial \alpha_i}\frac{1}{f}\frac{\partial f}{\partial \alpha_j} = -\frac{1}{f^2}\frac{\partial f}{\partial \alpha_j}\frac{\partial f}{\partial \alpha_i} + \frac{1}{f}\frac{\partial^2 f}{\partial \alpha_i \, \partial \alpha_j},$$

$$\mathbb{E}\left\{\frac{-\partial^2 \log f}{\partial \alpha_i \, \partial \alpha_j}\right\} = \int \frac{\partial \log f}{\partial \alpha_i}\frac{\partial \log f}{\partial \alpha_j} f \, dx - \int \frac{\partial^2 f}{\partial \alpha_i \, \partial \alpha_j} \, dx.$$

We note that

$$\int \frac{\partial^2 f}{\partial \alpha_i \, \partial \alpha_j} \, dx = \frac{\partial^2}{\partial \alpha_i \, \partial \alpha_j}\int f \, dx = \frac{\partial^2}{\partial \alpha_i \, \partial \alpha_j}(1) = 0,$$

$$(15.10)$$

$$(\Lambda^{-1})_{ij} = \mathbb{E}\left\{\frac{-\partial^2 w}{\partial \alpha_i \, \partial \alpha_j}\right\} = \frac{1}{2}E\left\{\frac{\partial^2 \chi^2}{\partial \alpha_i \, \partial \alpha_j}\right\}.$$

Equations (15.4), (15.9), and (15.11) provide three useful forms for obtaining $(\Lambda^{-1})_{ij}$. In principle, these equations should be evaluated at α_0, but in practice we evaluate them at α^*. We also sometimes find it convenient to replace the

$$\mathbb{E}\left\{\frac{\partial^2 w}{\partial \alpha_i \, \partial \alpha_j}\right\}$$

by the experimental result for $w = \log \mathcal{L}$ as a function of α_i and α_j. For a single parameter, these equations reduce to the useful results given in Chap. 12.

$$\text{variance } \alpha^* = \frac{1}{n}\left(\mathbb{E}\left\{\left(\frac{\partial \log f}{\partial \alpha}\right)^2\right\}\right)^{-1}$$

$$= \left(\mathbb{E}\left\{\left(\frac{\partial w}{\partial \alpha}\right)^2\right\}\right)^{-1} = \left(\mathbb{E}\left\{\frac{-\partial^2 w}{\partial \alpha^2}\right\}\right)^{-1}.$$

$$(15.11)$$

Again, we often use experimental values of w as a function of α in place of the expectation value.

15.2   Regression Analysis with Non-constant Variance

Suppose we have a set of independent measurements y_ν ($\nu = 1, \ldots, n$) at a set of parameter values $x_{1\nu}, x_{2\nu}, \ldots, x_{k\nu} = \mathbf{x}_\nu$. The y_ν each have a certain estimated variance s_ν^2. We define the weight of this point by

$$\omega_\nu = 1/s_\nu^2. \tag{15.12}$$

We assume here that the measured errors have approximately normal distributions. If this is not true, the method may be not be accurate. We will discuss this in Sect. 4 and in Chaps. 17 and 19 indicate more suitable methods if this problem occurs.

Very often we will consider that $\mathbf{x}$ is, in fact, a function expansion of a single parameter z. Then,

$$x_{j\nu} = \phi_j(z_\nu), \tag{15.13}$$

where the ϕ_j are a series of k independent functions. For example, we might have $x_{11} = z_1$, $x_{21} = z_1^2, \ldots,$.

For a concrete example of this, consider a problem in which we measure the values of a single phase shift at a series of energies, and wish to fit a theoretical curve to this. This example is illustrated in Fig. 15.1. Each point will have a different size error. We will consider the problem of handling measurements if there are errors in both the vertical and the horizontal positions of the measured points in Sect. 15.5.

Consider the important special case in which the dependence of y_ν on the parameters is linear, i.e., y_ν is a linear function of the $x_{j\nu}$. Define an approximating function $G(\mathbf{x})$ by

$$G(\mathbf{x}_\nu) = \sum_{j=1}^{k} \alpha_j x_{j\nu} = \sum_{j=1}^{k} \alpha_j \phi_j(z_\nu). \tag{15.14}$$

Fig. 15.1 Experimental points and a fitted theoretical curve.

ϕ_j might be a set of orthonormal functions, e.g., $\phi_j(z_v) = \sin(jz_v)$.

The set of $\alpha_1, \alpha_2, \ldots, \alpha_k$ are k arbitrary parameters whose values we will fit. Note that this is not a set of n identical trials, for each of the n points has a separate probability density. The trials are, however, independent. For our present situation, we have

$$(\Lambda^{-1})_{ij} = \sum_{p=1}^{n} \mathbb{E}\left\{ \frac{\partial \log f_p}{\partial \alpha_i} \frac{\partial \log f_p}{\partial \alpha_j} \right\}. \tag{15.15}$$

We use the modified least squares procedure, which we recall is equivalent to a maximum likelihood procedure. We wish to minimize

$$\chi^2 = \sum_{v=1}^{n} \frac{[y_v - G(\mathbf{x}_v)]^2}{\sigma_v^2} \cong \sum_{v=1}^{n} \omega_v [y_v - G(\mathbf{x}_v)]^2. \tag{15.16}$$

The approximation occurs since we have replaced $1/\sigma_v^2$ by its estimate ω_v. The resulting minimum of χ^2 will be asymptotically distributed in a χ^2 distribution with $n - k$ degrees of freedom if the distribution, in fact, has the assumed form. $\mathbb{E}\{y\} = G(x)$. Asymptotically, after fitting,

$$\mathbb{E}\{\chi^2\} = n - k. \tag{15.17}$$

Let us find the minimum χ^2.

$$\frac{\partial \chi^2}{\partial \alpha_s} = \sum_{v=1}^{n} -2\omega_v [y_v - G(\mathbf{x}_v)] x_{sv}. \tag{15.18}$$

This will be 0 at the minimum $G^*(x)$ corresponding to parameters α_j^*. At the minimum, we have

$$\sum_{v=1}^{n} \sum_{j=1}^{k} \alpha_j^* x_{jv} x_{sv} \omega_v = \sum_{v=1}^{n} \omega_v y_v x_{sv} \text{ for } s = 1, \ldots, k. \tag{15.19}$$

We can write this as

$$h\alpha^* = \mathbf{g},$$

$$\mathbf{g} = \text{vector} = \left\{ \sum_{v=1}^{n} \omega_v y_v x_{sv} \right\} = \left\{ \sum_{v=1}^{n} \omega_v y_v \phi_s(z_v) \right\}, \tag{15.20}$$

$$\boldsymbol{\alpha}^* = \text{vector} = \{\alpha_s^*\},$$

$$h = \text{matrix} = \left\{ \sum_{\nu=1}^{n} \omega_\nu x_{s\nu} x_{\ell\nu} \right\} = \left\{ \sum_{\nu=1}^{n} \omega_\nu \phi_s(z_\nu) \phi_\ell(z_\nu) \right\}.$$

We see that h is a symmetric $k \times k$ matrix and is independent of y_i, the experimental results. By inverting the matrix,

$$\alpha_s^* = \sum_{\ell=1}^{k} (h^{-1})_{s\ell} g_\ell. \tag{15.21}$$

Let us try to find the moment matrix Λ. Since h is independent of y_ν, the above equation tells us that $\boldsymbol{\alpha}^*$ is linear in y_ν. Thus, if the measurements are normally distributed, the distribution of $\boldsymbol{\alpha}^*$ is normal. If the expected value of $\boldsymbol{\alpha}$ is $\boldsymbol{\alpha}_0$, we recall that we have defined $\Lambda_{ij} = \mathbb{E}\{(\alpha_i^* - \alpha_{i0})(\alpha_j^* - \alpha_{j0})\}$. To estimate Λ, we use our previous result:

$$(\Lambda^{-1})_{ij} = \frac{1}{2} \mathbb{E} \left\{ \frac{\partial \chi^2}{\partial \alpha_i \, \partial \alpha_j} \right\} = \mathbb{E} \left\{ \frac{\partial(-\log \mathcal{L})}{\partial \alpha_i \, \partial \alpha_j} \right\}. \tag{15.22}$$

By taking another derivative of the equation for $\partial \chi^2 / \partial \alpha_s$, we have

$$(\Lambda^{-1})_{ij} \cong \mathbb{E} \left\{ \sum_{\nu=1}^{n} \omega_\nu x_{i\nu} x_{j\nu} \right\} = \mathbb{E}\{h_{ij}\} = h_{ij},$$

$$\Lambda_{ij} \cong (h^{-1})_{ij}. \tag{15.23}$$

This result is an approximation because we have used ω_ν, which are only estimates of $1/\sigma_\nu^2$. We have thus found the estimate $\boldsymbol{\alpha}^*$ and the error matrix Λ_{ij}.

If we are fitting a curve to a histogram, we must remember that we need to compare the integral of the density function over each bin to the contents of the bin. In practice people often, incorrectly, just use the value of the density function at the center of each bin for the comparison. This can lead to a biased result and an incorrect estimate of errors. We discussed an approximate correction in Chap. 13.

Next we consider the problem of the estimate of the error at an interpolated $\mathbf{x_p} = (x_{1p}, x_{2p}, \ldots, x_{kp})$ corresponding to functions ϕ_s evaluated at z_p.

$$\sigma_{z_p}^2 = \mathbb{E} \left\{ (G^*(\mathbf{x_p}) - \mathbb{E}\{y(\mathbf{x_p})\})^2 \right\}. \tag{15.24}$$

We have noted the solution for $\boldsymbol{\alpha}^*$ is linear in y_ν. Thus, the solution of the same form, but with the expected value of y_ν, will yield the expected values $\boldsymbol{\alpha}_0$ of the parameters.

Thus,

$$\sigma_{z_p}^2 = \mathbb{E}\left\{\sum_{\ell=1}^{k}\sum_{s=1}^{k}(\alpha_s^* - \alpha_{0s})x_{sp}(\alpha_\ell^* - \alpha_{0\ell})x_{\ell p}\right\}.$$

Using the definition of Λ_{ij} and the above equation for Λ_{ij}, we have

$$\sigma_{z_p}^2 = \sum_{\ell=1}^{k}\sum_{s=1}^{k}(h^{-1})_{\ell s}x_{sp}x_{\ell p}$$

$$= \sum_{\ell=1}^{k}\sum_{s=1}^{k}(h^{-1})_{\ell s}\phi_s(z_p)\phi_\ell(z_p). \tag{15.25}$$

We now compare the variance of $G^*(\mathbf{x})$ at the measured points with the original estimates of the measurement error at that point. Consider

$$\delta^2 = \frac{1}{n}\sum_{\nu=1}^{n}\sigma_{z_\nu}^2\omega_\nu.$$

$$\delta^2 = \frac{1}{n}\sum_{\nu=1}^{n}\sum_{\ell=1}^{k}\sum_{s=1}^{k}\omega_\nu x_{s\nu}x_{\ell\nu}(h^{-1})_{\ell s}$$

$$= \frac{1}{n}\sum_{\ell=1}^{k}\sum_{s=1}^{k}h_{s\ell}(h^{-1})_{\ell s}, \tag{15.26}$$

$$\delta^2 = \frac{k}{n}.$$

Remember that k is the number of parameters we fit and n is the number of experimental points in that space.

Note that $\delta^2 < 1$, since $k < n$. The average error of the fitting curve is less than the average error of observation since the curve uses many observations.

Let us examine the average error over an interval. We easily see that if $x_{i\nu} = \phi_i(z_\nu)$, we have

$$\overline{\sigma_z^2} \equiv \frac{1}{z_2 - z_1}\int_{z_1}^{z_2}\sigma_z^2\,dz,$$

$$\overline{\sigma_z^2} = \frac{1}{z_2 - z_1}\sum_{\ell=1}^{k}\sum_{s=1}^{k}(h^{-1})_{\ell s}\int_{z_1}^{z_2}\phi_s(z)\phi_\ell(z)\,dz. \tag{15.27}$$

Worked Problem 15.1 will provide an illustration of using the above methodology in practice.

Consider the application of this method to fitting histograms containing weighted events. If there are n events in a given bin, each with weight W_i, the bin is treated as if it held $\sum_{i=1}^{n} W_i$ events. Monte Carlo calculations, for example, often lead to weighted events. (The term "weights" here is a completely different concept than when used previously to mean the inverse of the variance in a term in a least squares sum.)

If we wish to fit histograms of weighted events, it is necessary to find the variance of the weighted bins. A single unweighted event has a standard deviation of $\sqrt{1} = 1$. If the event is multiplied by a constant W_i, the standard deviation is also multiplied by that constant and becomes W_i. Hence, the variance for a single weighted event is the square of the standard deviation W_i^2. The variance for the bin is $\sum_{i=1}^{n} W_i^2$. The fitting then proceeds in the normal manner.

Suppose we are fitting our observations to a series as above. Where can we terminate? Do we need term k or not? Let χ_k^2 be the minimum value of χ^2 if we fit using k terms. Suppose we do not need the kth term, i.e., suppose $\alpha_{ko} = 0$. Let

$$S_M = \frac{M}{\chi_k^2}(\chi_{k-1}^2 - \chi_k^2), \tag{15.28}$$

$$M = n - k.$$

The denominator of S_M has a χ^2 distribution in M degrees of freedom. The numerator can be shown (Cziffra and Moravscik 1958) to have a χ^2 distribution in one degree of freedom. Hence, S_M has the F distribution with one degree of freedom in the numerator and M degrees of freedom in the denominator. We can now test S_M to see if the extra term is necessary. This should always be used in conjunction with the test of χ^2. Furthermore, if term k is 0, term $k + 1$ may still not be 0. For example, all odd terms may vanish in a particular problem.

Another method of checking for termination consists of checking the value of a parameter estimate, α_j^*, and its variance. If the parameter estimate is non-zero at a specified confidence level, then it is taken as significant. of a signal For example, $\alpha_j^{*2}/\sigma_{\alpha_j^*}^2 > 3.84$ corresponds to a 95% confidence level limit. Both methods are often used. The first method above, using the χ^2 of the fit, is probably preferable.

15.3   The Gibbs Phenomenon

A problem can occur with these methods. If the functions $\phi_j(z)$ are an orthonormal set of functions, then cutting off abruptly after k terms may lead to a "Gibbs phenomenon", a high-frequency ringing especially at the borders of the region (Blobel 1984).

Let us examine this problem. We continue to assume we have a linear fit and, as above, we let

$$\chi^2 = \sum_{\nu=1}^{n}(y_\nu - \alpha_j x_{j\nu} - b_\nu)^2 \omega_\nu, \tag{15.29}$$

where y_ν ($\nu = 1, \ldots, n$) are measurements at a set of parameter values $x_{j\nu}$. ω_ν is the weight of point ν and the α_j are the parameters we are trying to fit [$x_{j\nu} = \phi_j(z_\nu)$]. We have introduced b_ν as a background function, assuming that y_ν is a sum of signal plus background. Including background, we now have

$$\mathbf{g} = \text{vector} = \left\{ \sum_{\nu=1}^{n} \omega_\nu (y_\nu - b_\nu) x_{s\nu} \right\} = \left\{ \sum_{\nu=1}^{n} \omega_\nu (y_\nu - b_\nu) \phi_s(z_\nu) \right\}, \quad (15.30)$$

$$h = \Lambda_{\exp}^{-1} = \text{matrix} = \left\{ \sum_{\nu=1}^{n} \omega_\nu x_{s\nu} x_{l\nu} \right\} = \left\{ \sum_{\nu=1}^{n} \omega_\nu \phi_s(z_\nu) \phi_l(z_\nu) \right\}, \quad (15.31)$$

$$\chi^2 = \text{constant} - 2\boldsymbol{\alpha}^T \mathbf{g} + \boldsymbol{\alpha}^T \Lambda_{\exp}^{-1} \boldsymbol{\alpha}. \quad (15.32)$$

We now proceed to find the minimum using essentially the method worked out in Sect. 15.1. However, we will modify the procedure slightly to better illustrate the present problem.

Since $\Lambda_{\exp}^{-1}$ is symmetric and positive definite, it can be transformed to a diagonal matrix D by an orthogonal transformation. The eigenvalues will be approximately $1/\sigma_i^2$. We will call the eigenvectors u_i. Let the orthogonal transformation matrix be U_1.

$$D = U_1^T \Lambda_{\exp}^{-1} U_1. \quad (15.33)$$

We arrange the eigenvalues in order of decreasing values ($1/\sigma_1^2 \geq 1/\sigma_2^2 \cdots$). Typically the variances vary over several orders of magnitude. Let $D^{1/2}$ be the diagonal matrix with $D_{ii}^{1/2} = 1/\sigma_i$. Now use these matrices to transform our parameters to a new set:

$$\boldsymbol{\alpha} = U_1 D^{-1/2} \mathbf{a}. \quad (15.34)$$

In this new basis $\Lambda_{\exp}^{-1}$ is a unit matrix. Inserting this expression into our expression above for χ^2 and ignoring constant terms, we obtain

$$\chi^2(\mathbf{a}) = -2\mathbf{a}^T D^{-1/2} U_1^T \mathbf{g} + \mathbf{a}^T \mathbf{a}. \quad (15.35)$$

We want the gradient of χ^2 to be 0 for the minimum. Thus, we obtain

$$\mathbf{a}^* = D^{-1/2} U_1^T \mathbf{g}. \quad (15.36)$$

The covariance matrix of the solution vector $\mathbf{a}^*$ is the unit matrix by our construction. We can now transform back to our original parameters, obtaining

$$\boldsymbol{\alpha}_{\text{unreg}}^* = \sum_{s=1}^{k} \sigma_s a_s u_s. \quad (15.37)$$

The problem here is the weighting by σ_s. This means that the most insignificant term, i.e., the one with the largest uncertainty, may get a large weight. Suppose $a_s = \sin(s\theta)$. The high orders oscillate very quickly. Even in general cases, the higher-order functions are likely to have high-frequency oscillations, and the fitted function then tends to oscillate. The nature of the problem is now clear.

The solution to the oscillation problem lies in not cutting off the higher terms abruptly, but in introducing a gentle cutoff of higher-order terms.

15.4   The Regularization Method

We saw in the last section that cutting off the series abruptly could give a high weight to the badly determined higher-order terms. Regularization methods provide a way of cutting off these terms gradually.

The local curvature of a function $G(x)$ is

$$\frac{G''(x)}{[1 + (G'(x))^2]^{3/2}} \approx G''(x) \quad \text{if } |G'(x)| << 1.$$

Therefore, we take $r = \int_a^b [G''(x)]^2 \, dx$ as an approximate measure of the smoothness of a function over the interval (a, b). A rapidly oscillating function will have large local curvatures and a large value for r. Instead of minimizing χ^2, we will minimize the modified function:

$$\chi^2_{\text{reg}} = \chi^2 + \tau r(\boldsymbol{\alpha}). \tag{15.38}$$

This is known as the Tikhonov regularization method and τ is known as the regularization parameter. (Tikhonov 1963; Tikhonov and Arsenin 1977; Phillips 1962) The extra term introduces a penalty for rapid oscillation.

To second order in the parameters, let $r(\alpha) = \alpha^T C \alpha$, where C is a symmetric matrix. (We will suppose there is no term linear in the parameters.) If we again make the transformation U_1, we now have

$$\chi^2_{\text{reg}} = -2\mathbf{a}^T D^{-1/2} U_1^T \mathbf{g} + \mathbf{a}^T \mathbf{a} + \tau \mathbf{a}^T D^{-1/2} U_1^T C U_1 D^{-1/2} \mathbf{a}. \tag{15.39}$$

This last term can be written as

$$\tau \mathbf{a}^T C_1 \mathbf{a}, \tag{15.40}$$

where

$$C_1 = D^{-1/2} U_1^T C U_1 D^{-1/2}. \tag{15.41}$$

C_1 can be transformed to a diagonal matrix L by an orthogonal transform U_2. Thus, $L = U_2^T C_1 U_2$. We can adjust U_2 so that the eigenvalues of L are in increasing order.

Define $\mathbf{a} = U_2 \mathbf{a}'$. Since this is a pure rotation, we have $\tau \mathbf{a}^T C_1 \mathbf{a} = \tau (\mathbf{a}')^T L \mathbf{a}'$ and we have

$$\chi^2_{\text{reg}}(\mathbf{a}') = -2(\mathbf{a}')^T U_2^T D^{-1/2} U_1^T \mathbf{g} + (\mathbf{a}')^T (I + \tau L)\mathbf{a}'. \tag{15.42}$$

Here I is a unit matrix. Again, we find the minimum by setting the gradient of $\chi^2_{\text{reg}} = 0$.

$$(\mathbf{a}')^*_{\text{reg}} = (I + \tau L)^{-1} U_2^T D^{-1/2} U_1^T \mathbf{g} = (I + \tau L)^{-1} (\mathbf{a}')^*_{\text{unreg}}. \tag{15.43}$$

We have noted here that the unregularized solution corresponds to setting $\tau = 0$. Transforming back to the original coefficients α, we have

$$\boldsymbol{\alpha}^*_{\text{reg}} = U_1 D^{-1/2} U_2 (\mathbf{a}')^*_{\text{reg}}, \qquad \boldsymbol{\alpha}^*_{\text{unreg}} = U_1 D^{-1/2} U_2 (\mathbf{a}')^*_{\text{unreg}}. \tag{15.44}$$

Alternatively, we can consider the parameters a' of a set of transformed basis functions $\phi'_j(z)$, which are linear combinations of the original $\phi_i(z)$:

$$\boldsymbol{\phi}'(z) = U_2^T D^{-1/2} U_1^T \boldsymbol{\phi}(z). \tag{15.45}$$

Suppose we decide that after m_0, the coefficients are not significant. With the unregularized method, we have m_0 terms. For the regularized method, we have

$$(a'_j)^*_{\text{reg}} = \frac{1}{1 + \tau L_{jj}} (a'_j)^*_{\text{unreg}}. \tag{15.46}$$

Note that we do not have to redo the fit. An advantage of performing this diagonalization is that we only need to divide each unregularized coefficient by the $1 + \tau L_{jj}$ factor. For $\tau L_{jj} \ll 1$, the factor $(1 + \tau L_{jj})$ is close to one. As we go to higher terms, the L_{jj} factor typically increases rapidly. A useful choice is to choose τ such that

$$m_0 = \sum_{j=1}^{m} \frac{1}{1 + \tau L_{jj}}. \tag{15.47}$$

Here m_0 is to be chosen big enough so that significant coefficients are not attenuated severely. Often this means choosing m_0 just a little past the value obtained from significance tests. Other choices are discussed in the next Section.

The covariance matrix of $(\mathbf{a}')^*_{\text{unreg}}$ is the unit matrix and that of $(\mathbf{a}')^*_{\text{reg}}$ is

$$\text{cov}[(\mathbf{a}')^*_{\text{reg}}] = (I + \tau L)^{-2}. \tag{15.48}$$

This can easily be transformed back to find the covariance matrix for α^*.

15.5   Other Regularization Schemes

We next indicate some other choices for r, i.e., some other regularization techniques for imposing smoothness on the data. The first of these other techniques is called the principle of maximum entropy or MaxEnt (Shannon and Weaver 1949). Note that the number of ways of putting N objects into k bins with n_i in bin i is given by the multinomial coefficient $N!/(n_1! n_2! \cdots n_k!)$, described in Worked Problem 5.1. Assume that N is fixed and use Stirling's approximation for each factorial. The log of this coefficient is then, up to an additive constant,

$$N(\log N - 1) - \sum_{i=1}^{k} n_i(\log n_i - 1) = -\sum_{i=1}^{k} n_i \log \frac{n_i}{N} + \text{constant.}$$

This can be viewed as an entropy and that is the motivation for this method. We let

$$r = \sum_{i=1}^{k} \frac{n_i}{N} \log \frac{n_i}{N}; \quad \chi_{\text{reg}}^2 = \chi^2 + \tau r.$$

This entropy is maximum if all of the n_i are equal.

A variant of this method can be used if it is believed that a good approximation to the true distribution is known. Let q_i be the probability of an event falling into bin i using this assumed approximation, and let the true distribution probability (which is to be found) be p_i. Then the cost function, cross-entropy (Kullback 1977; Jaynes 1968) is given by

$$r(p, q) = \sum_{i=1}^{k} p_i \log \frac{p_i}{k q_i}; \quad \chi_{\text{reg}}^2 = \chi^2 + \tau r(p, q). \tag{15.49}$$

These methods will be discussed further in Chap. 20 in connection with training neural networks. With any of these smoothing functions, a criterion for choosing τ is needed. We list below several possibilities that have been proposed.

1. Choose τ such that the expected error in the bias is approximately the size of the statistical error.
2. Choose τ such that $\chi^2 \approx 1$ per bin.
3. Choose τ to minimize $\sum_{i=1}^{m}(\sigma_{i\,\text{est}}^2 + b_i^2)$, where $\sigma_{i\,\text{est}}^2$ is the estimated statistical variance in bin i, and b_i is the estimated bias.
4. Choose τ such that $\chi_{\text{bias}}^2 \equiv \sum_{i=1}^{m} b_i^2 \approx m$, the number of bins.

It may be necessary to try, using Monte Carlo methods, various of these possibilities for r and for τ for each different experimental situation. Different methods will be best in different situations. Some further discussion and examples are given in Cowan (1998). Still another prescription for gently cutting off the higher-order terms has been given by Cutkosky (1969, 1972). He suggests that an improvement can be obtained in some cases by the use of a convergence test function, ψ. $\chi^2 + \psi$ is minimized. ψ is chosen to have an approximate χ^2 distribution and to automatically provide a slow cutoff for higher-order terms if the correlation of the α_s^* terms is small.

Let

$$R_k = \frac{k|\alpha_k|^2}{\sum_{s=0}^{k-1}|\alpha_s|^2},$$

$$\rho_k = k \, \log\left\{1 + \frac{R_k}{k}\left(1 - \frac{0.58}{k}\right)\right\}, \tag{15.50}$$

$$\psi = \sum \rho_k.$$

This last function is an empirical approximation of a much more complicated form. To simplify the minimization, Cutkosky suggests that α_s be determined sequentially up to high s values.

At each stage, the sum in the denominator of R_k is to be held at the previously determined value. Empirically, this has given good results in a series of tests, far better than the sharp cutoff in the series described above. Further details are given in the Cutkosky article.

15.6 Fitting Data with Errors in Both x and y

In the previous sections, we have discussed measuring the value of y_ν at a set of specific values x_ν. From these data, an estimate was made of a set of k parameters α_j, $j = 1, k$, in a function $y(x; \alpha)$. However, often we are faced with a problem in which, at each point, both x and y are measured variables and both have experimental errors.

Assume that the errors in x and y are normally distributed, are known, and are independent. Let the true values at each point be $x_{0\nu}$, $y_{0\nu}$. Then $w \equiv$ the log of the likelihood function is

$$w = \sum_{\nu=1}^{n}\left[-\frac{1}{2}\log 2\pi\sigma_{x_\nu}^2 - \frac{1}{2}\log 2\pi\sigma_{y_\nu}^2 - \frac{(x_\nu - x_{0\nu})^2}{2\sigma_{x_\nu}^2} - \frac{(y_\nu - y_{0\nu})^2}{2\sigma_{y_\nu}^2}\right].$$

Ignore the first two terms, which are independent of the parameters and the data.

$$\chi^2 = -2w = \sum_{\nu=1}^{n}\left[\frac{(x_\nu - x_{0\nu})^2}{\sigma_{x_\nu}^2} + \frac{(y_\nu - y_{0\nu})^2}{\sigma_{y_\nu}^2}\right]. \tag{15.51}$$

Because both x and y are individually random variables, we have $2n$ measured variables. There are k parameters α_j and n parameters $x_{0\nu}$. Note that given the true

α_j and the $x_{0\nu}$, the $y_{0\nu}$ are determined from $y_{0\nu} = y(x_{0\nu})$. The number of degrees of freedom is then $2n - n - k = n - k$, the same value as found for the problem with fixed x values. We find the minimum in the usual way:

$$\frac{\partial \chi^2}{\partial x_{0\nu}} = 0 = -2\frac{(x_\nu - x_{0\nu})}{\sigma_{x_\nu}^2} - 2\left[\frac{(y_\nu - y_{0\nu})}{\sigma_{y_\nu}^2}\right]\frac{dy}{dx}\bigg|_{x_{0\nu}}.$$

This is an exact set of equations and can be solved by iteration. Often, this set of equations can be simplified. The technique is known as the effective variance method (Orear 1982; Barker and Diana 1974). Make a linear approximation for $y(x)$ near $x_{0\nu}$, $y(x_\nu) = y(x_{0\nu}) + y'(x_{0\nu})(x_\nu - x_{0\nu})$, where $y'(x_{0\nu}) \equiv dy/dx\big|_{x_\nu=x_{0\nu}}$. Note that $y_{0\nu} \equiv y(x_{0\nu}) = y(x_\nu) - y'(x_{0\nu})(x_\nu - x_{0\nu})$. Again set

$$\frac{\partial \chi^2}{\partial x_{0\nu}} = 0 = -2\frac{(x_\nu - x_{0\nu})}{\sigma_{x_\nu}^2} - 2\left\{\frac{y_\nu - [y(x_\nu) - y'(x_{0\nu})(x_\nu - x_{0\nu})]}{\sigma_{y_\nu}^2}\right\}y'(x_{0\nu}).$$

Then

$$(x_\nu - x_{0\nu})\left(\frac{1}{\sigma_{x_\nu}^2} + \frac{[y'(x_{0\nu})]^2}{\sigma_{y_\nu}^2}\right) + \frac{[y_\nu - y(x_\nu)]y'(x_{0\nu})}{\sigma_{y_\nu}^2} = 0.$$

$$x_\nu - x_{0\nu} = -\frac{[y_\nu - y(x_\nu)]y'(x_{0\nu})}{\sigma_{y_\nu}^2/\sigma_{x_\nu}^2 + [y'(x_{0\nu})]^2} = -\frac{\sigma_{x_\nu}^2 y'(x_{0\nu})[y_\nu - y(x_\nu)]}{\sigma_{y_\nu}^2 + [y'(x_{0\nu})]^2\sigma_{x_\nu}^2}.$$

Define

$$\sigma_\nu^2 \equiv \sigma_{y_\nu}^2 + [y'(x_{0\nu})]^2\sigma_{x_\nu}^2. \tag{15.52}$$

Then, at the minimum with respect to $x_0\nu$ of χ^2,

$$\chi^2_{\text{min}-x_{0\nu}} = \sum_{\nu=1}^{n}\frac{[y'(x_{0\nu})(\sigma_{x_\nu}^2/\sigma_\nu^2)(y_\nu - y(x_\nu))]^2}{\sigma_{x_\nu}^2}$$

$$+ \sum_{\nu=1}^{n}\frac{\{y_\nu - (y(x_\nu) + y'(x_{0\nu})[\sigma_{x_\nu}^2/\sigma_\nu^2]y'(x_{0\nu})[y_\nu - y(x_\nu)])\}^2}{\sigma_{y_\nu}^2}$$

$$= \sum_{\nu=1}^{n}\left([y'(x_{0\nu})]^2[y_\nu - y(x_\nu)]^2\frac{\sigma_{x_\nu}^2}{\sigma_\nu^4}\right) \tag{15.53}$$

$$+ \frac{(\{y_\nu - y(x_\nu)\}\{1 - [y'(x_{0\nu})]^2\sigma_{x_\nu}^2/\sigma_\nu^2\})^2}{\sigma_{y_\nu}^2}.$$

After a little algebra, this equation can be reduced to

$$\chi^2_{\min-x_{0\nu}} = \sum_{\nu=1}^{n} \frac{[y_\nu - y(x_\nu)]^2}{\sigma_\nu^2}. \qquad (15.54)$$

This can then be minimized with respect to y in the usual way. Hence, in this approximation, one simply uses the methodology from the previous sections with $\sigma_{x\nu}^2$ replaced by σ_ν^2. Start by guessing values for dy/dx, solving for the parameters, and iterating. Usually two iterations are sufficient. If the x and y errors are correlated, one can rotate coordinates at each point to obtain independent variables and then apply this method. See Chap. 16 for an alternative treatment of this problem.

15.7 Non-linear Parameters

Suppose next that our parameters are not linear. For example, we might wish to fit

$$G(x) = \alpha_1 \, \sin(x + \alpha_2) \quad \text{or} \quad Ae^{-\lambda_1 t} + Be^{-\lambda_2 t}. \qquad (15.55)$$

From the general results shown in Chap. 12, we can still minimize using χ^2 as defined by

$$\chi^2 \cong \sum_{\nu=1}^{n} \omega_\nu [y_\nu - G(\mathbf{x}_\nu)]^2. \qquad (15.56)$$

χ^2 is still asymptotically distributed according to a χ^2 distribution in $n - k$ degrees of freedom as long as the errors in y_ν are approximately normal. In many cases, the moment matrix is best found by using

$$(\Lambda^{-1})_{ij} = \frac{1}{2}\mathbb{E}\left\{\frac{\partial \chi^2}{\partial \alpha_i \, \partial \alpha_j}\right\} \cong \frac{1}{2}\frac{\partial \chi^2 \, \text{experimental}}{\partial \alpha_i \, \partial \alpha_j}. \qquad (15.57)$$

The F test described above is still valid.

The non-linear case is a very common one and it is often very difficult to find the minimum analytically. I will try to indicate below some of the issues involved in trying to find the function minimum and in trying to interpret the errors. I will not try to explore here the very sophisticated algorithms that have been developed to attack this difficult problem. I suggest that you read James (1972a, b), James and Roos (1972).

1. A minimum can be a stationary point, a cusp, or the edge of a region as indicated below (Figs. 15.2, 15.3 and 15.4).
2. When a minimum is found, it is often not obvious whether it is a local or a global minimum. See Fig. 15.5.
3. Suppose the function is quadratic in the parameters we are varying, as happens for the χ^2 function for linear parameters as indicated above. We have seen

Fig. 15.2 Example of a
minimum being at a
stationary point

Fig. 15.3 Example of a
minimum being at a cusp

Fig. 15.4 Example of a
minimum being at the edge
of a region

Fig. 15.5 Example of a
local minimum different
from the global minimum

that for this example the first and second derivatives have all of the information we need and we can calculate the minimum position in one step. Usually for non-linear parameters, we must estimate the minimum by an iteration scheme.

However, the first and second derivative information is usually used in some way even in these schemes.

4. There is no "generally" superior algorithm for iteration. The programs we use are often of two kinds. The first is a general program, such as the CERN program MINUIT, which has several iteration schemes that can be selected. This kind is useful if you are going to minimize your function only a few times and need to see what works. The second application occurs if you have a specific problem to be redone many times. An example of this occurs in particle physics when we have measured the momenta of the incoming and outgoing particles in a particle interaction and wish to see how well the interaction may fit various hypotheses. In some experiments, we may have 10^6 such interactions to analyze. Here, we tailor an algorithm for that particular problem and use a specialized program optimizing everything we can about that specific problem.

5. We would like to minimize the number of times we have to evaluate the function to be minimized. If we are taking the first and second derivatives numerically, then, for k parameters, each evaluation of the second derivative involves $k(k - 1)/2$ evaluations of the function in addition to the ones needed to obtain the gradient.

6. If we are in a region where the function is varying almost linearly, we must take care that the step size does not diverge.

7. We must check that we are finding a minimum not a maximum. For several variables, we, therefore, must check that the second derivative matrix, G, is positive definite. If it is not, as often happens, we must derive ways to handle it.

8. We must take care that the method does not oscillate around the minimum without approaching it quickly.

9. If two variables are highly correlated, the minimization is difficult. Suppose in an extreme case two variables appear only as a sum, $(a + b)$. Varying a and b, but keeping the sum constant, will not change the function value and the program will not have a unique minimum point.

10. Many of the methods involve matrix inversion or its equivalent and often double precision is needed.

11. In some programs (e.g., MINUIT), if one attempts to set limits to the values of parameters that can be searched, the fitting process can be seriously slowed down.

12. Several of the simplest methods have serious drawbacks:

Searches just taking a grid of points are usually impractical, especially for multi-variable problems.

If we vary only one of the parameters at a time or use the direction of steepest descent each time, then successive steps necessarily go in orthogonal directions and are inefficient. A narrow valley can cause very slow convergence. See Fig. 15.6.

Fig. 15.6 Varying only one parameter at a time can lead to very slow convergence

We might pretend the function is approximately quadratic and use the values of the first and second derivatives at the present point to estimate the next step. This method is often unstable.

The modern methods that are popular now are very sophisticated, some involving varying the metric of the spaces. A review of some of these methods is given in Cousins and Highland (1992) and Cousins (1995).

After a minimum is found, interpretation of the errors quoted also needs some discussion. For χ^2 functions, one standard deviation corresponds to $\chi^2 - \chi^2_{min} = 1$. For the negative logarithm of the maximum likelihood, one standard deviation corresponds to $w_{max} - w = 0.5$. Because χ^2 is quadratic in the errors, two standard deviations corresponds to $\chi^2 - \chi^2_{min} = 4$, and three corresponds to the difference being 9, etc. One can estimate the errors by assuming that near the minimum, the function is approximately quadratic. Then, the diagonal elements of the error matrix ($\equiv$ moment or covariance matrix $= \Lambda = G^{-1}$) are the (σ^2)'s including the effects of correlation, i.e., the total σ's. The diagonal terms of the second derivative matrix, G, of course, do not include the correlations. The correlations appear with the inversion.

A more accurate way of calculating the errors is to trace the contours of constant χ^2. If one has two parameters, α_1 and α_2, then Fig. 15.7 shows the relation between the one standard deviation contour and the one standard deviation limit on a single parameter. (The contour shown would be a pure ellipse for a quadratic function.) To find the overall limits on a single parameter, α_i, you do not just change the value of that one parameter, but you must re-minimize the function at each step with respect to all of the other parameters, corresponding to the new value of α_i.

The probability of being within a hypervolume with $\chi^2 - \chi^2_{min} = C$ decreases (for fixed C) as the number of parameters, N, increases. You can just read this off from a χ^2 distribution table as it is just the probability of exceeding the value C for the

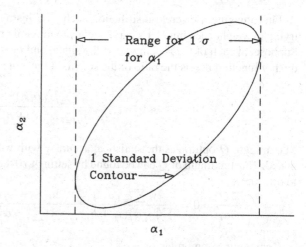

Fig. 15.7 Relation between a one standard deviation limit and a one standard deviation contour for a two-parameter problem

Table 15.1 C for Multiparameters.: The confidence Level (probability that the contents are within a hypercontour of $\chi^2 = \chi^2_{min} + C$)

No. Param.ǀ C =	50%	70%	90%	95%	99%
1	0.46	1.07	2.70	3.84	6.63
2	1.39	2.41	4.61	5.99	9.21
3	2.37	3.67	6.25	7.82	11.36
4	3.36	4.88	7.78	9.49	13.28
5	4.35	6.06	9.24	11.07	15.09
6	5.35	7.23	10.65	12.59	16.81
7	6.35	8.38	12.02	14.07	18.49
8	7.34	9.52	13.36	15.51	20.09
9	8.34	10.66	14.68	16.92	21.67
10	9.34	11.78	15.99	18.31	23.21
11	10.34	12.88	17.29	19.68	24.71

χ^2 function with N degrees of freedom. Table 15.1 (taken from James 1972b) gives some useful values. Confusion about this point has produced errors in numerous experimental articles in print.

15.8 Optimizing a Data Set with Signal and Background

A common problem is that of trying to fit data that have both signal and background. Often parts of the data set are mostly background. Data cuts are placed to select the region in which the signal is most significant. Here, we will examine one simple method of trying to evaluate where cuts might be placed. We follow closely the treatment of Akerlof (1989).

First imagine a discrete distribution such as a histogram. We will proceed by trying to weight each bin i by a parameter α_i to maximize the ratio of signal to error. Suppose, at each of the n points, s_i is the signal and b_i is the background. Define a quality function that is the ratio of the square of the signal divided by the variance.

$$Q(\alpha) \equiv \frac{S^2}{V} \equiv \frac{(\sum_{i=1}^{n} \alpha_i s_i)^2}{\sum_{j=1}^{n} \alpha_j^2 (s_j + b_j)}. \tag{15.58}$$

Maximizing Q optimizes the statistical accuracy with which the signal can be estimated. The maximum of Q can be found by setting to 0 the n derivatives with respect to α_i.

$$\frac{\partial Q}{\partial \alpha_i} = 0 = \frac{2s_i \sum_{k=1}^{n} \alpha_k s_k}{\sum_{j=1}^{n} \alpha_j^2 (s_j + b_j)} - \frac{2\alpha_i (s_i + b_i)(\sum_{k=1}^{n} \alpha_k s_k)^2}{[\sum_{j=1}^{n} \alpha_j^2 (s_j + b_j)]^2}.$$

Removing common factors,

$$0 = s_i \sum_{j=1}^{n} \alpha_j^2 (s_j + b_j) - \alpha_i (s_i + b_i) \sum_{j=1}^{n} \alpha_j s_j.$$

The solutions can be seen to be given by

$$\alpha_i = \frac{s_i}{s_i + b_i}, \quad Q_{\max} = \sum_{i=1}^{n} \frac{s_i^2}{s_i + b_i}. \tag{15.59}$$

If the distribution is continuous, then $s_i \rightarrow s(\lambda)$, $b_i \rightarrow b(\lambda)$, and $\alpha_i \rightarrow \alpha(\lambda)$. Then the maximum is given by

$$\alpha(\lambda) = \frac{s(\lambda)}{s(\lambda) + b(\lambda)}, \quad Q_{\max} = \int \frac{s(\lambda)^2 d\lambda}{s(\lambda) + b(\lambda)}. \tag{15.60}$$

Consider, as an example of this technique, the problem of identifying a faint photon signal from an X-ray star in the presence of a much larger, isotropic background. Since $s(\lambda) << b(\lambda)$,

$$\alpha(\lambda) \approx s(\lambda)/b(\lambda), \quad Q_{\max} \approx \int \frac{s(\lambda)^2 d\lambda}{b(\lambda)}.$$

Assume the error in determining the direction of a given photon is normally distributed in both projections with variance σ^2 on each projection and that $\theta^2 = \theta_x^2 + \theta_y^2$. Here the variable is the angle θ. Then,

$$s(\theta) d\theta_x d\theta_y = \frac{1}{2\pi\sigma^2} e^{-\theta^2/2\sigma^2} d\theta_x d\theta_y.$$

$$Q_{\max} = \frac{1}{4\pi^2 b\sigma^4} \int_0^\infty e^{-\theta^2/2\sigma^2} 2\pi\theta \, d\theta = \frac{1}{4\pi b\sigma^2}.$$

A slight variation in the assumptions can make a significant difference in the conclusions. Consider the problem in astronomy of counting photons from a wide band source of light in the presence of background. The total photon rate is $s(\lambda) + b(\lambda)$, which is integrated over some range of wavelengths. Suppose one wishes to design an optimized optical filter to maximize the signal significance. The photons, after passing the filter, are counted, but their wavelength is not measured. Using notation similar to the last problem,

$$S(\alpha) = \int_{\lambda_1}^{\lambda_2} \alpha s \, d\lambda.$$

The variance is now different from before. Only $\alpha(s + b)d\lambda$ photons are detected at each λ. Since the standard deviation is the square root of that number, the variance is

$$V(\alpha) = \int_{\lambda_1}^{\lambda_2} \alpha(s + b) \, d\lambda.$$

$$Q(\alpha) = \frac{\left(\int_{\lambda_1}^{\lambda_2} \alpha s \, d\lambda\right)^2}{\int_{\lambda_1}^{\lambda_2} \alpha(s + b) \, d\lambda}. \tag{15.61}$$

Again one looks at the solution for $\alpha(\lambda)$. One can show that, unlike the previous problem, at each wavelength, the optimum weight is either 0 or 1, i.e., the filter must either be totally transparent or totally absorbing. As before, one can easily modify to the problem of a set of discrete physical bins, if desired. For the continuous problem, one can search for the maximum Q by including wavelengths with successively smaller values of $s(\lambda)/b(\lambda)$.

Consider a problem similar to the last example. Suppose we wish to design an angular aperture for a detector accepting a collimated beam of photons in a uniform background. Here, again, the variable is the angle. Let the angular distribution of the beam be

$$s(\theta) = \frac{1}{\sqrt{2\pi\sigma^2}} e^{-\theta^2/2\sigma^2}.$$

Assume the aperture is to be set with a cutoff parameter θ_0 and that the signal is small compared to the background b. Q is given by

$$Q = \frac{\left([1/(2\pi\sigma^2)]^2 \int_0^{\theta_0} e^{-\theta^2/2\sigma^2} 2\pi\theta \, d\theta\right)^2}{\pi\theta_0^2 b}.$$

Let $u = (\theta_0/\sigma)^2$. Solving $\partial Q/\partial\theta_0 = 0$ gives

$$(1 + u)e^{-u/2} - 1 = 0, \quad \left(\frac{\theta_0}{\sigma}\right)^2 = u \approx 2.51286.$$

15.9   Robustness of Estimates

The methods discussed here and in the previous chapter are very powerful if the underlying distributions are close to normal. If, for example, we are fitting to a histogram, the individual bins have a Poisson distribution that is close to normal if the number of elements in a bin is not too small. However, if we are fitting to a curve and the errors in each point are not close to normal, e.g., have large tails, then the estimates of goodness of fit may be seriously biased.

An estimate is **robust** if it is relatively independent of the underlying distribution. It turns out that estimates of parameters based on the means of distributions, such as those based on Student's distribution, are relatively robust. However, those based on variances, such as the least squares test, are quite sensitive to non-normal errors. They are especially sensitive to distributions with non-zero kurtosis. For example, suppose a result for a χ^2 distribution with five degrees of freedom corresponds to a 5% level if we, incorrectly, assume normally distributed errors when, in fact, there is a non-zero kurtosis. Then, if the kurtosis is -1, the level is in fact 0.0008. If the kurtosis is 1, the level is 0.176, and if the kurtosis is 2, the level is 0.315.

When we run into this kind of distribution, we can attempt to find new variables that are more normally distributed or we can try to find tests less sensitive to the underlying distribution. In Chap. 17, we consider the first of these methods, and, in Chap. 19, we consider the second method.

One method to see if the problem might be occurring is to plot the distributions of the individual terms in the χ^2 sum and see whether they are distributed in an approximately normal distribution or whether they are skewed or have a non-zero kurtosis. If the distribution is close to normal, then the method is probably satisfactory, but you should turn to one of the methods we will discuss later if the distribution is significantly different from the normal distribution.

In this chapter, we have examined methods for fitting data to curves, estimating several parameters. For the instance of linear parameters, the problem can be reduced to a crank-turning method and this is illustrated in a worked problem. The non-linear case is intrinsically harder, but often can be handled with the aid of a computer program which can find the minimum of a function in multidimensional space. In the Exercises, we will look at some tractable examples of this type of problem.

15.10   Worked Problems

WP15.1 In neutrino physics one measures a parameter y. Consider the reaction $v + N \rightarrow \mu + X$. We define $y = (E_v - E_\mu)/E_\mu$. From this distribution for each

E_ν, it is possible to derive $B = (Q - \overline{Q})/(Q + \overline{Q})$, where $(1 - B)/2$ is a measure of the fraction of antiquarks in the nucleus.

Values of B for $\overline{\nu}$ in the energy range 10–50 GeV are given in Table 15.2 (the last two columns refer to the fit to be described below).

Assume that the measurement errors are approximately normal and fit these data to the form:

$$B = \alpha_1 + \alpha_2 \overline{E}.$$

Answer:

$$\text{Let } \mathbf{g}_{\frac{1}{2}} = \sum_{\nu=1}^{9} \omega_\nu y_\nu \phi_{\frac{1}{2}}(z_\nu),$$

where

$$\phi_1 = 1, \, \phi_2 = E,$$
$$y = B, \, z = E, \, \omega = 1/\sigma^2.$$

$$\mathbf{g}_{\frac{1}{2}} = \sum_{\nu=1}^{9} \frac{B}{\sigma^2} \begin{pmatrix} 1 \\ E \end{pmatrix} = \begin{pmatrix} 3017.5 \\ 100182.36 \end{pmatrix},$$

$$h = \sum_{\nu=1}^{9} \omega_\nu \phi_s(z_\nu)\phi_l(z_\nu)$$

$$= \begin{pmatrix} \sum_{\nu=1}^{9} \frac{1}{\sigma^2} & \sum \frac{E}{\sigma^2} \\ \sum \frac{E}{\sigma^2} & \sum \frac{E^2}{\sigma^2} \end{pmatrix} = \begin{pmatrix} 3805.6 & 1.2856 \times 10^5 \\ 1.2856 \times 10^5 & 4.874 \times 10^6 \end{pmatrix}.$$

Table 15.2 Experimental values of B versus $\overline{E}$

E_{range}	$\overline{E}$(GeV)	B	Values with fit	χ^2
30–40	35	0.85 ± 0.05	0.79	1.44
40–50	45	0.78 ± 0.05	0.75	0.36
50–60	55	0.68 ± 0.05	0.72	0.64
10–20	15	0.85 ± 0.05	0.85	0
20–30	25	0.80 ± 0.04	0.82	0.25
30–40	35	0.79 ± 0.04	0.79	0
40–50	43	0.78 ± 0.06	0.76	0.11
–	20	0.83 ± 0.05	0.84	0.04
–	40	0.76 ± 0.06	0.77	0.03
				2.87

- The cofactor$_{ji}$ is the determinant of the matrix with the jth row and ith column removed $\times (-1)^{i+j}$. The determinant of h is 2.0229×10^9. Thus,

$$(h^{-1})_{ij} = \frac{\text{cofactor}_{ji}}{(\text{determinant of } h)},$$

$$(h^{-1})_{ij} = \begin{pmatrix} 2.4095 \times 10^{-3} & -6.355 \times 10^{-5} \\ -6.355 \times 10^{-5} & 1.811 \times 10^{-6} \end{pmatrix}.$$

$$\alpha_1^* = \sum_{\ell=1}^{2} (h^{-1})_{1\ell}\, \ell g_\ell,$$

$$\alpha_1^* = 2.4095 \times 10^{-3} \times 3017.5 - 6.355 \times 10^{-5} \times 1.0018 \times 10^5$$

$$= 0.9042,$$

$$\alpha_2^* = -6.355 \times 10^{-5} \times 3017.5 + 1.881 \times 10^{-6} \times 1.0018 \times 10^5$$

$$= -3.323 \times 10^{-3}.$$

$$\sigma_{\alpha_1}^2 = \Lambda_{11} = (h^{-1})_{11} = 2.409 \times 10^{-3} \implies \sigma_{\alpha_1} = 4.9 \times 10^{-2},$$

$$\sigma_{\alpha_2}^2 = \Lambda_{22} = (h^{-1})_{22} = 1.88 \times 10^{-6} \implies \sigma_{\alpha_2} = 1.37 \times 10^{-3}.$$

$$\overline{(\alpha_1 - \alpha_1^*)(\alpha_2 - \alpha_2^*)} = -6.355 \times 10^{-5}.$$

$$B = (0.904 \pm 0.05) - (3.3 \pm 1.4) \times 10^{-3} E$$

$$\chi^2 = 2.87 \quad \text{for } 9 - 2 = 7 \text{ d.f.}$$

- This is an excellent fit. About 90% of time, we would have a larger χ^2. (If χ^2 is too low, this can also indicate problems in the fit. Perhaps the error estimate on the measurements was too conservative.)
- The expected error of the curve at a given interpolated point (if the form is right) is

$$\sigma = \sum_{\ell=1}^{2} \sum_{s=1}^{2} (h^{-1})_{\ell s} \phi_\ell(E) \phi_s(E)$$

$$= (h^{-1})_{11} + 2(h^{-1})_{12} E + (h^{-1})_{22} E^2.$$

- Consider 35 GeV, for example:

$$\sigma^2_{35\ GeV} = 2.4095 \times 10^{-3} - 2 \times 6.355 \times 10^{-5} \times 35$$
$$+ 1.881 \times 10^{-6} \times 50^2$$
$$= 2.652 \times 10^{-4} \Longrightarrow \sigma_{35\ GeV} = 0.016.$$

- Similarly $\sigma_{10\ GeV} = 0.025$, $\sigma_{50\ GeV} = 0.036$. Therefore, the errors are best determined in the center of the region. (Extrapolations are *much* less certain than interpolations especially if high powers are involved.)

- Note: For histograms, we force the solution to have the same area as the experiment. If that is not the case a maximum likelihood estimate must be made.

15.11 Exercises

15.1 The decay rate of orthopositronium has been measured by atomic physicists. Gidley and Nico (1991) have kindly provided me with a set of data in which they measured the number of decays in 10-nsec bins every 50-nsec, starting at 300-nsec. They have a background as well as a signal and fit to the form

$$R = Ae^{-\lambda t} + B. \tag{15.62}$$

The data are given in Table 15.3.

In Appendix A, a fit of parameters A, B, and λ to the data is done using MINUIT and the output is given there using the data for this exercise, as a demonstration of how MINUIT works. From looking at that output, find the estimated errors in these quantities. Suppose you would try to predict the rate that would have been measured at 525 ns. Find the predicted rate and the error.

Table 15.3 Number of decays in 10-nsec nins as a function of time

Region(t ns)	N	$\sqrt{N}$
300	803,000	896
350	581,083	762
400	429,666	655
450	320,016	566
500	242,783	493
550	188,487	434
600	150,737	388
650	124,103	352
700	105,397	325
750	92,748	305

15.2 You will again need a general minimizing program for this exercise. In Exercise 8.1, you worked out a method for generating Monte Carlo events with a Breit–Wigner (B-W) distribution in the presence of a background.   Here we will use a somewhat simpler background. Suppose the background is proportional to $a + bE$ with $a = b = 1$. The B–W resonance has energy and width (E_0, Γ). The number of events in the resonance is 0.22 times the total number of events over the energy range 0–2. We further set:

$$E_0 = 1,$$
$$\Gamma = 0.2.$$

Generate 600 events. Save the energies on a file for use in calculating maximum likelihood, and for making a histogram for the next problem. Ignore any B–W points with energy past 2. (Regenerate the point if the energy is past 2.) Now suppose that you are trying to fit the resultant distribution of events to a B–W distribution and a background term of the above form with b, E_0, Γ to be determined from your data. Take a to be the actual value 1, and introduce a new parameter $fract$, the fraction of events in the B–W. The likelihood for each event will be

$$f = fract \times f_{BW} + (1 - fract) \times f_{\text{back}},$$

where both density functions are properly normalized. Using the maximum likelihood method, fit the events you have generated and see how closely the fitted parameters correspond to the actual ones. You will have to take account here that for the parameters given, only about 93% of the B–W is within the interval 0–2 GeV. As noted in Sect. 15.7, you need to recalibrate the errors since 1 standard deviation corresponds to $w_{max} - w = 0.5$ rather than the value of 1.0 for a χ^2 fit. Directions for doing this for MINUIT are given in the example MINUIT fit in Appendix A.

15.3 Try the above exercise, first taking the file saved in Exercise 15.2 and binning the data into a 40 bin histogram. Then use the χ^2 method with a general minimizing program. Although some information is lost by binning, the fitting is often simpler and less time-consuming on the computer when the data are binned. Each evaluation of χ^2 for this method involves evaluation of the density function only N_{bin} times not N_{event} times as was the case for the maximum likelihood method. Your bins should be considerably narrower than the width Γ of the resonance if you wish to get accurate results. For this exercise, you may use the value of the density function at the center of the bin times the bin width rather than the exact integral over the bin. (As we indicated earlier, for careful work, you should not use this approximation.) Plot the data histogram and the fits from Exercise 15.2 and this exercise.

References

Akerlof C (1989) Private communication, University of Michigan, Michigan

Barker D, Diana L (1974) Am J Phys 42:224

Blobel V (1984) Unfolding methods in high-energy physics experiments. Technical report, DESY, D2000, Hamburg-52, Germany

Cousins R (1995) Am J Phys 63:398

Cousins R, Highland V (1992) Nuc Inst Meth A320:331

Cowan G (1998) Statistical data analysis. Oxford Clarendon Press, Oxford

Cutkosky R (1972) Technical report, Carnegie-Mellon University, Pittsburgh, PA 15213

Cutkosky R (1969) Theory of representation of scattering data by analytic functions. Ann Phys 54:350

Cziffra P, Moravscik M (1958) A practical guide to the method of least squares. Technical report, Lawrence Berkeley Laboratory

Gidley D, Nico J (1991) Private communication

James F (1972a) Function minimization. In: Proceedings of the 1972 CERN computing and data processing school, Pertisau, Austria, 10-24 September, 1972

James F (1972b) Interpretation of the errors on parameters as given by minuit. Technical report, CERN-DD division, CERN, CH-1211, Geneva, 23 Switzerland

James F, Roos M (1972) Minuit, function minimization and error analysis. Technical report, CERN-DD division, CERN, CH-1211, Geneva, 23 Switzerland

Jaynes E (1968) Prior probabilities. IEEE Trans Syst Sci Cybern SSC- 4:227

Kullback S (1977) Information theory and statistics. Wiley, New York

Orear J (1982) Am J Phys 50:912

Phillips D (1962) A technique for the numerical solution of certain integral equations of the first kind. J ACM 9:84

Shannon C, Weaver W (1949) The mathematical theory of communication. University of Illinois Press, Urbana

Tikhonov A (1963) On the solution of improperly posed problems and the method of regularization. SovMath 5:1035

Tikhonov A, Arsenin V (1977) Solutions of Ill-posed problems. Wiley, New York

Chapter 16
Fitting Data with Correlations and Constraints

Abstract Until now, we have usually taken individual measurements x as independent but this is often not the case (Böck 1961). The measurements may have correlations between them. Also, in addition to the parameters of interest θ, the results may depend on the various measurement and other parameters we label *nuisance parameters* v. The probability is then $P_x(x|\theta, v)$. These additional parameters may reduce systematic effects, but they increase statistical error. In addition, there may be correlations between the nuisance variables and the variables of interest. Ambiguities and bias in the result for the parameters of interest can then occur and the effective number of degrees of freedom in the problem might be altered (Roe 2013). Methods for handling these nuisance variables are described. A general approximate linear fitting method is discussed (somewhat similar to the gradient descent method to be discussed in Chap. 20). This method leads to the definition of "pulls" or "stretch variables" which are quite useful in seeing if a fit looks reasonable. Finally, methods of putting together estimates of parameters from several experiments are discussed.

16.1 Handling Nuisance Variables

It is sometimes possible to reduce the impact of nuisance variables by doing additional measurements to constrain their value, giving data y (Cowan 2018). Suppose these measurements are statistically independent from x. and have probability $P_y(y|v)$. The likelihood then becomes

$$\mathcal{L}(\theta| v) = P_x(x|\theta, v)P_y(y|v).$$

For Monte Carlo simulation, we usually want a test statistic to be independent of the nuisance variables. Sometimes we may want to average over v using $P_y(v)$ although this may spread out the distribution, since nature only uses the true nuisance values. The profile likelihood helps avoid this problem

$$\mathcal{L}_p(\theta) = \mathcal{L}(\theta, v_{\max}), \tag{16.1}$$

© Springer Nature Switzerland AG 2020
B. P. Roe, *Probability and Statistics in the Physical Sciences*,
Undergraduate Texts in Physics,
https://doi.org/10.1007/978-3-030-53694-7_16

where ν_{max} is the value of the vector which maximizes $\mathcal{L}$ for the given θ. We want to construct a test statistic q_θ which increases as the incompatibility of the theory and data increases. This can be done approximately by using the profile likelihood if we use the likelihood ratio of the profile likelihood divided by the likelihood using the values $\theta_{overall\ max}$, $\nu_{overall\ max}$ which maximize the likelihood

$$R = \frac{\mathcal{L}(\theta, \nu_{max})}{\mathcal{L}(\theta_{overall\ max}, \nu_{overall\ max})} \tag{16.2}$$

As it did in our previous discussion of likelihood ratios, Wilkes theorem (Wilks 1946) states that $-2\ln R(\theta)$ approaches a χ^2 distribution as the size of the data sample gets very large. For a histogram of n bins, the number of degrees of freedom is n minus the size of the vector θ.

16.2   Constraints on Nuisance Variables

Furthermore, there may be constraints on the values. We will examine here a general formalism for dealing with these constraints if the problem can be approximately linearized and if the errors on each point are approximately normal.

We will consider the general formalism briefly, and then discuss some examples in which these features occur.

Suppose we measure m variables $x = x_1, x_2, \ldots, x_m$, which may be correlated, and whose true values are x_0. Here

$$\chi^2 = \sum_{j=1}^{m} \sum_{i=1}^{m} (x_{0i} - x_i)(\Lambda^{-1})_{ij}(x_{0j} - x_j), \tag{16.3}$$

where Λ is the moment matrix.

If there are also c constraint equations, we must modify the formalism to include them. The variables in the constraint equations include nuisance variables. Let the constraint equations be

$$f^k(y, x) = 0, \quad k = 1, c. \tag{16.4}$$

y_i ($i = 1, u$) are u unknown parameters. We use the method of Lagrange multipliers. We let our multipliers be p_k, $k = 1, c$. Then

$$\chi^2 = \sum_{j=1}^{m} \sum_{i=1}^{m} (x_{0i} - x_i)(\Lambda^{-1})_{ij}(x_{0j} - x_j) + 2\sum_{k=1}^{c} p_k f^k(y, x). \tag{16.5}$$

This just adds 0 to χ^2 if the equations of constraint are obeyed. In matrix notation, we have

$$\chi^2 = (x_0 - x)^T \Lambda^{-1}(x_0 - x) + 2p^T f. \tag{16.6}$$

Fig. 16.1 Example of a problem in which x, z pairs having correlated errors are measured at several points along a curve

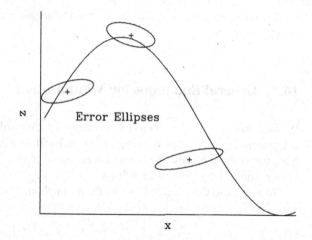

Error Ellipses

Here T means transpose. x and f are column vectors.

Where do we use the formalism described in this section? We will give two examples.

In Sect. 15.5, we considered the problem of fitting events, where the events were functions of two variables and there were measurement errors in both variables. This problem can also be handled with the present formalism. Suppose we are making a set of measurements (x, z) of points on one or more curves, but both x and z are uncertain. For instance, we might measure the emissivity (z) of a body in a given frequency range as a function of temperature (x) under conditions in which both the emissivity and temperature measurements have significant errors.

If there are n measurements and we rename the (x, z) pairs as (x_i, x_{i+1}), then the χ^2 is as given in Eq. 16.3, with $m = 2n$. If the measurements are independent, then Λ is a set of 2×2 matrices on the diagonal, giving the error matrix for each (x, z) pair (Fig. 16.1). However, we have not yet put in the fact that we want these points to be on the curve(s). To do this, we must add n equations of constraint, one for each measurement pair: $f^k(x_i, x_{i+1}) = 0$. The index k labels the different measurement pairs. If we are fitting just one curve, then all f's are the same function.

Another instance of this method occurs if we are trying to fit constrained events. Imagine we have some sort of tracking device in which we are measuring the process, $\pi^- p \to \pi^- p$. Suppose we know the incoming beam momentum and measure the outgoing momentum of each particle. We ask to improve our measurements by using the four equations of constraint at the interaction point corresponding to the conservation of momentum (three equations) and energy (one equation). We can also check that the data are consistent with this reaction and distinguish against, for example, $\pi^- p \to \pi^- \pi^0 p$. The u parameters here may correspond to unmeasured variables, such as the momentum components of the π^0.

The measurements are often correlated depending on the specific properties of the measuring apparatus. For example, if momentum is measured by curvature of the particle path in a magnetic field, then the initial angles and the curvature are usually strongly correlated.

16.3 General Equations for Minimization

We will now look at the method of solution. $\partial \chi^2 / \partial x$ will be a vector with the ith component $\partial \chi^2 / \partial x_i$. f_x will mean $\partial f / \partial x$ and will be a $c \times m$ matrix (not square). Note that if there is only one constraint equation, $c = 1$; f_x is a row, not a column vector. Similarly, f_y will mean $\partial f / \partial y$.

We again use the modified minimum χ^2 method. For a minimum (at $x_0 = x^*$), we require

$$\frac{\partial \chi^2}{\partial x} = 2[(x^* - \backslash x)^T \Lambda^{-1} + p^{*T} f_x] = 0,$$

$$\frac{\partial \chi^2}{\partial y} = 2p^{*T} f_y = 0, \qquad (16.7)$$

$$\frac{\partial \chi^2}{\partial p} = 2f_x = 0.$$

The last equation shows that this solution satisfies the constraint equations (Eq. 16.4). Usually, the set of Eq. 16.7, are very difficult to solve analytically. One of the general minimization programs (such as (MINUIT) is often used. However, as we've seen in Chap. 13, the total number of events in the histogram must not be a parameter.

A useful relation, true in the linear approximation, which we will derive in the next section, is that

$$\frac{(x_i - x_i^*)}{\sqrt{\sigma_{i,\text{unfitted}}^2 - \sigma_{i,\text{fitted}}^2}} \qquad (16.8)$$

is normal with mean 0 and variance 1. Here x_i^* is the value of x_i at the minimum and $\sigma_{i,\text{fitted}}^2$ is the variance of x_i^*. This variable is called a "pull" or a "stretch" variable and is an exceedingly useful test in practice to check if initial measurements errors are assigned correctly and if there are biases.

16.4 Iterations and Correlation Matrices

In the problem of fitting constrained events, we often have many thousands or even millions of events and cannot afford to use the very general fitting programs for an event-by-event fit. Usually, we have to arrange the events into histograms. However, sometimes we have to devise one that suits our particular problem and works reasonably quickly. We outline here one of the standard methods of iteration (Böck

1960, 1961; Ronne 1968). In practice, one needs to add procedures to handle cases in which the results do not converge. The formulas in this section look somewhat cumbersome, but are easily handled by computers. I have included derivations since it is hard to find them in standard texts.

We start off with a zeroth-order solution obtained usually by ignoring all but u of the constraints. This enables us to get starting values for the u parameters. We then proceed to iterate. We assume that the functions are smooth and the changes small.

Let v be the index of the iteration. We will find estimates x^*, p^*, y^* for step $v + 1$ using derivatives evaluated using step v. We want $f = 0$. At step v, f will not quite be 0. We set

$$f^v + f_x^v(x^{*v+1} - x^{*v}) + f_y^v(y^{*v+1} - y^{*v}) = 0. \tag{16.9}$$

Thus, we change parameters so that in this linearized approximation, they drive f to 0. This will fix the iteration step. Using $\Lambda^T = \Lambda$, Eq. 16.7, tell us

$$x^{*v+1} = x - \Lambda f_x^{vT} p^{*v+1}. \tag{16.10}$$

Remember x with no superscript is the measured value of x. Let

$$R = f^v + f_x^v(x - x^{*v}),$$
$$S = f_x^v \Lambda f_x^{vT}. \tag{16.11}$$

Note that R and S only depend on step v. S is a square $c \times c$ matrix. From Eqs. (16.9) and (16.11), we have

$$R + f_x^v(x^{*v+1} - x) + f_y^v(y^{*v+1} - y^{*v}) = 0. \tag{16.12}$$

We then use Eq. (16.10), to obtain

$$R - f_x^v(\Lambda f_x^{vT} p^{*v+1}) + f_y^v(y^{*v+1} - y^{*v}) = 0. \tag{16.13}$$

Next, use Eq. (16.11)

$$p^{*v+1} = S^{-1}[R + f_y^v(y^{*v+1} - y^{*v})]. \tag{16.14}$$

We use this result in the transpose of the second line of Eq. (16.7), $2f_y^T p = 0$.

$$f_y^{vT} S^{-1} R + f_y^{vT} S^{-1} f_y^v(y^{*v+1} - y^{*v}) = 0,$$

$$y^{*v+1} = y^{*v} - (f_y^{vT} S^{-1} f_y^v)^{-1} f_y^{vT} S^{-1} R. \tag{16.15}$$

We can now perform one step of the iteration. We use Eq. (16.15) to obtain $y^{*\nu+1}$. Using this, we obtain $p^{*\nu+1}$ from Eq. (16.14). Finally, using Eq. (16.10), we obtain $x^{*\nu+1}$.

To find χ^2 at step $\nu + 1$, we start with Eq. (16.3), which does not have the equations of constraint attached. x_0 is to be replaced by its estimate $x^{*\nu+1}$ using Eq. (16.10).

$$
\begin{aligned}
\chi^2 &= (-\Lambda f_x^{\nu T} p^{*\nu+1})^T \Lambda^{-1}(x^{*\nu+1} - x) \\
&= (p^{*\nu+1})^T[-f_x^{\nu}(x^{*\nu+1} - x)] \\
&= (p^{*\nu+1})^T[f_x^{\nu}(x - x^{*\nu}) - f_x^{\nu}(x^{*\nu+1} - x^{*\nu})].
\end{aligned}
\tag{16.16}
$$

We used $\Lambda^T = \Lambda$ in the above. Now use Eqs. (16.9) and (16.11)

$$
\begin{aligned}
\chi^2 &= (p^{*\nu+1})^T[R - f^{\nu} + f^{\nu} + f_y^{\nu}(y^{*\nu+1} - y^{*\nu})], \\
\chi^2 &= (p^{*\nu+1})^T[R + f_y^{\nu}(y^{*\nu+1} - y^{*\nu})].
\end{aligned}
\tag{16.17}
$$

We now have the basic iteration procedure. We can go from step ν to $\nu + 1$ and evaluate the new χ^2.

Next, we wish to examine how we can estimate the errors in our fitted values. $x^{*\nu+1}$, $y^{*\nu+1}$ are implicit functions of the measured variables x, and we wish to propagate errors. We use a first-order approximation for the propagation. Suppose

$$
x^{*\nu+1} = g(x),
$$

$$
\tag{16.18}
$$

$$
y^{*\nu+1} = h(x),
$$

$$
\mathbb{E}\left\{(x_i^{*\nu+1} - x_{i0})(x_j^{*\nu+1} - x_{j0})\right\} \approx \sum_{k,\ell} \frac{\partial g_i}{\partial x_k} \frac{\partial g_j}{\partial x_\ell} \mathbb{E}\left\{(x_k - x_{k0})(x_\ell - x_{\ell 0})\right\}.
$$

We will call the first expectation value (above) $(\Lambda_g)_{ij}$ and the second is the moment matrix $\Lambda_{k\ell}$. Thus,

$$
\Lambda_{x^*} \equiv \Lambda_g = \frac{\partial g}{\partial x} \Lambda \left(\frac{\partial g}{\partial x}\right)^T.
\tag{16.19}
$$

This formula is generally useful when propagating errors; we are changing variables. In a similar manner, we obtain $\Lambda_{y^{*\nu+1}}$ and the correlation matrix between correlation matrix measured variables and the parameters:

$$
\Lambda_{y^{*\nu+1}} \equiv \Lambda_h = \frac{\partial h}{\partial x} \Lambda \left(\frac{\partial h}{\partial x}\right)^T,
\tag{16.20}
$$

$$
\Lambda_{x^{*\nu+1}, y^{*\nu+1}} = \frac{\partial g}{\partial x} \Lambda \left(\frac{\partial h}{\partial x}\right)^T.
\tag{16.21}
$$

We note that $\Lambda_{x^{*v+1},\, y^{*v+1}}$ is not a symmetric matrix. The correlation between x_2^* and y_5^* bears no relation to the correlation between x_5^* and y_2^*. Thus,

$$\Lambda_{y^{*v+1},\, x^{*v+1}} \equiv \Lambda_{x^{*v+1},\, y^{*v+1}}^T \neq \Lambda_{x^{*v+1},\, y^{*v+1}}. \qquad (16.22)$$

In fact, $\Lambda_{x^{*v+1},\, y^{*v+1}}$ is not generally a square matrix but is an $m \times u$ rectangular matrix.

We must now find $\partial g / \partial x$ and $\partial h / \partial x$. We linearize the equations by taking f_x^v as fixed. From Eq. (16.10), we have

$$\frac{\partial g}{\partial x} \equiv \frac{\partial x^{*v+1}}{\partial x} = 1 - \Lambda f_x^{vT} \frac{\partial p^{*v+1}}{\partial x}.$$

We substitute for p^{*v+1} using Eq. (16.14)

$$\frac{\partial g}{\partial x} = 1 - \Lambda f_x^{vT} S^{-1} \left[\frac{\partial R}{\partial x} + f_y^v \partial \frac{(y^{*v+1} - y^{*v})}{\partial x} \right].$$

Next, we use Eq. (16.15)

$$\frac{\partial g}{\partial x} = 1 - \Lambda f_x^{vT} S^{-1} \left(\frac{\partial R}{\partial x} - f_y^v \left(f_y^{vT} S^{-1} f_y^v \right)^{-1} f_y^{vT} S^{-1} \frac{\partial R}{\partial x} \right).$$

From Eq. (16.11)

$$\frac{\partial R}{\partial x} = \frac{\partial}{\partial x} [f^v + f_x^v (x - x^{*v})].$$

By f_x^v, we mean

$$f_x^v = \frac{\partial f(x^{*v})}{\partial x^{*v}},$$

$$\frac{\partial}{\partial x} f^v = \frac{\partial f(x^{*v})}{\partial x} = f_x^v \frac{\partial x^{*v}}{\partial x},$$

$$\frac{\partial R}{\partial x} = \left[f_x^v \frac{\partial x^{*v}}{\partial x} + f_x^v \left(1 - \frac{\partial x^{*v}}{\partial x} \right) \right] = f_x^v.$$

We have again taken the derivatives constant in the differentiations. Thus,

$$\frac{\partial g}{\partial x} = 1 - \Lambda f_x^{vT} S^{-1} \left(f_x^v - f_y^v \left(f_y^{vT} S^{-1} f_y^v \right)^{-1} f_y^{vT} S^{-1} f_x^v \right).$$

Using the above considerations, we also have

$$\frac{\partial h}{\partial x} = - \left(f_y^{vT} S^{-1} f_y^{vT} \right)^{-1} f_y^{vT} S^{-1} f_x^v.$$

This is the derivative $\partial(y^{*\nu+1} - y^{*\nu})/\partial x$. In our present linearized formalism with derivatives kept fixed, we wish to fix $y^{*\nu}$ as well. Let

$$K = f_y^{\nu T} S^{-1} f_y^{\nu}. \tag{16.23}$$

Then

$$\frac{\partial g}{\partial x} = 1 - \Lambda f_x^{\nu T} S^{-1}\left(f_x^{\nu} - f_y^{\nu} K^{-1} f_y^{\nu T} S^{-1} f_x^{\nu}\right),$$

$$\frac{\partial h}{\partial x} = -K^{-1} f_y^{\nu T} S^{-1} f_x^{\nu}. \tag{16.24}$$

The moment matrices can now be found from Eqs. (16.19)–(16.21)

$$
\begin{aligned}
\Lambda_{x^{*\nu+1}} &= \left[1 - \Lambda f_x^{\nu T} S^{-1}\left(f_x^{\nu} - f_y^{\nu} K^{-1} f_y^{\nu T} S^{-1} f_x^{\nu}\right)\right] \\
&\quad \times \Lambda\left[1 - \Lambda f_x^{\nu T} S^{-1}\left(f_x^{\nu} - f_y^{\nu} K^{-1} f_y^{\nu T} S^{-1} f_x^{\nu}\right)\right]^T \\
&= \Lambda - \Lambda f_x^{\nu T} S^{-1}\left(f_x^{\nu} - f_y^{\nu} K^{-1} f_y^{\nu T} S^{-1} f_x^{\nu}\right)\Lambda \\
&\quad - \Lambda\left(f_x^{\nu T} - f_x^{\nu T}\left(S^{-1}\right)^T f_y^{\nu}\left(K^{-1}\right)^T f_y^{\nu T}\right)\left(S^{-1}\right)^T f_x^{\nu}\Lambda \\
&\quad + \Lambda f_x^{\nu T} S^{-1}\left(f_x^{\nu} - f_y^{\nu} K^{-1} f_y^{\nu T} S^{-1} f_x^{\nu}\right) \\
&\quad \times \Lambda\left(f_x^{\nu T} - f_x^{\nu T}\left(S^{-1}\right)^T f_y^{\nu}\left(K^{-1}\right)^T f_y^{\nu T}\right)\left(S^{-1}\right)^T f_x^{\nu}\Lambda \\
&= \Lambda - \Lambda f_x^{\nu T} S^{-1}\left(1 - f_y^{\nu} K^{-1} f_y^{\nu T} S^{-1}\right)f_x^{\nu}\Lambda \\
&\quad - \Lambda f_x^{\nu T}\left(1 - \left(S^{-1}\right)^T f_y^{\nu}\left(K^{-1}\right)^T f_y^{\nu T}\right)\left(S^{-1}\right)^T f_x^{\nu}\Lambda \\
&\quad + \Lambda f_x^{\nu T} S^{-1}\left(S - f_y^{\nu} K^{-1} f_y^{\nu T}\right) \\
&\quad \times \left[1 - \left(S^{-1}\right)^T f_y^{\nu}\left(K^{-1}\right)^T f_y^{\nu T}\right]\left(S^{-1}\right)^T f_x^{\nu}\Lambda.
\end{aligned} \tag{16.25}
$$

Since Λ, S, and K are all symmetric, the last term becomes

$$\Lambda f_x^{\nu T}\left(1 - S^{-1} f_y^{\nu} K^{-1} f_y^{\nu T}\right)\left(1 - S^{-1} f_y^{\nu} K^{-1} f_n^{\nu T}\right) S^{-1} f_x^{\nu}\Lambda. \tag{16.26}$$

Note that $U = f_y^{\nu} K^{-1} f_y^{\nu T} S$ has the property that $U^2 = U$. Writing out all the terms and canceling, we then get

$$\Lambda_{x^{*\nu+1}} = \Lambda - \Lambda f_x^{\nu T} S^{-1} f_x^{\nu}\Lambda + \Lambda f_x^{\nu T} S^{-1} f_y^{\nu} K^{-1} f_y^{\nu T} S^{-1} f_x^{\nu}\Lambda, \tag{16.27}$$

$$\Lambda_{y^{*\nu+1}} = K^{-1} f_y^{\nu T} S^{-1} f_x^{\nu}\Lambda f_x^{\nu T} S^{-1} f_y^{\nu} K^{-1},$$

$$\Lambda_{y^{*\nu+1}} = K^{-1} = \left(f_y^{\nu T} S^{-1} f_y^{\nu}\right)^{-1}. \tag{16.28}$$

Another moment matrix of considerable interest is $\Lambda_{x^{*\nu+1},x}$, the correlation matrix between the fitted and measured values of x.

$$\Lambda_{x^{*\nu+1},x} = \frac{\partial g}{\partial x}\Lambda,$$

$$\tag{16.29}$$

$$\Lambda_{x^{*\nu+1},x} = \left[1 - \Lambda f_x^{\nu T} S^{-1}\left(f_x^{\nu} - f_y^{\nu} K^{-1} f_y^{\nu T} S^{-1} f_x^{\nu}\right)\right]\Lambda.$$

Therefore

$$\Lambda_{x^{*\nu+1},x} = \Lambda - \Lambda f_x^{\nu T} S^{-1} f_x^{\nu}\Lambda + \Lambda f_x^{\nu T} S^{-1} f_y^{\nu} K^{-1} f_y^{\nu T} S^{-1} f_x^{\nu}\Lambda,$$

$$\tag{16.30}$$

$$\Lambda_{x^{*\nu+1},x} = \Lambda_{x^{*\nu+1}}.$$

Consider

$$\begin{aligned}
\text{variance}\left(x_i^{\nu+1} - x_i\right) &= \mathbb{E}\left\{\left(x_i^{*\nu+1} - x_{i0} + x_{i0} - x\right)^2\right\} \\
&= \text{variance}\left(x_i^{*\nu+1}\right) + \text{variance}\left(x_i\right) \\
&\quad + 2\mathbb{E}\left\{\left(x_i^* - x_{i0}\right)\left(x_{i0} - x\right)\right\} \\
&= \left(\Lambda_{x^{*\nu+1}}\right)_{ii} + \Lambda_{ii} - 2\left(\Lambda_{x^{*\nu+1},x}\right)_{ii}.
\end{aligned}$$

$$\tag{16.31}$$

Using Eq. (16.30), we have

$$\begin{aligned}
\text{variance}\left(x_i^{*\nu+1} - x_i\right) &= \left(\Lambda - \Lambda_{x^{*\nu+1}}\right)_{ii} \\
&= \sigma_{i,\text{unfitted}}^2 - \sigma_{i,\text{fitted}}^2.
\end{aligned}$$

$$\tag{16.32}$$

Hence, we have proved that the "pull",

$$\frac{x_i - x_i^*}{\sqrt{\sigma_{i,\text{unfitted}}^2 - \sigma_{i,\text{fitted}}^2}}$$

$$\tag{16.33}$$

indeed has mean 0 and variance 1. If, when plotted, the pulls are not normally distributed, that is an indication that either the errors are badly estimated or that the errors are inherently non-normal. In this latter case, as was discussed in Chap. 15, the goodness of fit is unreliable. To use this formalism, some methodology, such as will be described in Chap. 17 must be used to make the errors approximately normal. Chapter 19 will discuss some methods of estimation if we are unable to do this.

$$\begin{aligned}
\Lambda_{x^{*\nu+1},\, y^{*\nu+1}} &= -\left(1 - \Lambda f_x^{\nu T} S^{-1}\left(f_x^{\nu} - f_y^{\nu} K^{-1} f_y^{\nu T} S^{-1} f_x^{\nu}\right)\right) \\
&\quad \times \Lambda f_x^{\nu T} S^{-1} f_y^{\nu} K^{-1},
\end{aligned}$$

$$\tag{16.34}$$

$$\Lambda_{x^{*\nu+1},\, y^{*\nu+1}} = -\Lambda f_x^{T} S^{-1} f_y^{\nu} K^{-1}.$$

$$\tag{16.35}$$

The actual fitting process involves a great deal of sophistication in determining where to stop iterating, when the steps are not converging, and what to do if they do not converge. Badly measured variables require special handling because some of the matrices become almost singular. Further problems involve converging to a local minimum or a saddle point rather than the "best" minimum (if indeed the function has a minimum). See the discussion in Chap. 15. It is difficult to recognize these false minima and hard to set a general procedure for all cases. Various empirical solutions are used. Most computer programs also have facilities for cut steps. That is, if an iteration seems to walk away from or past the solution, the program has facilities to retake this iteration using a smaller step size. Setting up your own program for situations involving fitting many events requires considerable care, and considerable testing. The method described here is similar to the method of gradient descent which will be discussed in Chap. 20. For many problems, the very advanced minimization programs such as MINUIT, perhaps using Lagrange multipliers, are preferable to this step-by-step program.

We have examined here a general formalism for dealing with problems in which the individual measurements are correlated and/or there are constraints on the values. The formalism is an iterative one depending on an approximate linearization of the problem. As we have indicated, this can be a complicated procedure in practice and care must be taken in its application.

16.5   Putting Together Several Probability Estimates

In the previous section, we discussed estimating a vector of parameters x and the errors in the estimates from a single experiment. Often you have the task of estimating the parameters given the estimates from several experiments. For example, you may have looked at the angular distribution of scattered particles and at the overall rate of scattering separately. You may have several devices examining particles in a particle beam to see if they are π or k particles, each device giving a probability. You may have several measurements of a physical quantity (e.g., α) each with its own non-gaussian errors. For definiteness, I will discuss the first problem above.

If the measurement of probability for the angular distribution used the χ^2 or maximum likelihood methods, then the rate compared to the expected rate can be added in as simply another term. However, in a more general case, this may not be possible. One might, for example, use the Smirnov–Cramèr–Von Mises goodness of fit test to be discussed in Chap. 19 for the angular distribution and simply use Poisson distribution statistics to compare the rate with the expected rate.

We discuss below an analytic method for combining estimates (James 2006). Proofs of some of the steps will be left for the exercises. Suppose we have done n experiments to find a vector of parameters x, with dimension d each of which found x_i. The error matrix for the sum of the experiments is Λ, while the error matrix for an individual experiment i is Λ_i. For the moment we assume the individual experiments are independent of each other. Make a least squares fit to find the best fit value x_f. Let

$$\chi^2 = \sum_{i=1}^{n} (x_i - x)^T \Lambda_i^{-1}(x_i - x), \text{ where } \Lambda_i^{-1}$$

is the inverse of the error matrix Λ_i. We wish to find

$$\frac{\partial \chi^2(x)}{\partial x}\Big|_{x=x_f} = 0$$

The solution is

$$x_f = \left(\sum_{i=1}^{n} \Lambda_i^{-1} \right)^{-1} \cdot \left(\sum_{i=1}^{n} \Lambda_i^{-1} x_i \right) \tag{16.36}$$

with variance

$$\sigma^2 = \left(\sum_{i=1}^{n} \sigma_i^{-2} \right)^{-1}. \tag{16.37}$$

For $d = 1$, this simplifies to the simple weighted squares method discussed in Chap. 3,

$$x_f = \frac{\sum_{i=1}^{n} \sigma_i^{-2} x_i}{\sum_{i=1}^{n} \sigma_i^{-2}}.$$

Suppose we wish to look at one experiment k and see how compatible it is with the totality of all experiments, i.e., let

$$D_k = (x_f - x_k^T)\sigma_D^{-2}(x_f - x_k). \tag{16.38}$$

For σ_D^2 we get If there is no correlation between experiments then

$$(\sigma_D^2)_k = \Lambda_k - \Lambda \tag{16.39}$$

$$D_k = (x_f - x_k)^T (\Lambda_k - \Lambda)^{-1}(x_f - x_k) \tag{16.40}$$

Equation (16.40) also holds if there are correlations between experiments Finally, if x is one-dimensional we obtain

$$D_k = \frac{x_f - x_k)^2}{\sigma_k^2 - \sigma^2} \tag{16.41}$$

We again obtain the pull of experiment k on x_f to be

$$\text{pull}_k = \frac{x_f - x_k}{\sqrt{\sigma_k^2 - \sigma^2}} \tag{16.42}$$

with a normal distribution (0,1).

Below, I discuss another method I have found to be useful for putting independent results together.

Suppose that for each result, we obtain an integral probability level. If the data, in fact, are from the assumed distribution, then the probability levels (likelihoods) are themselves distributed with a flat distribution, i.e., 5% of the time one will get 5% or less as the probability level. The method starts with the probability levels of the individual results. This eliminates the disparities in the methods used to obtain them: Smirnov–Cramèr–Von Mises in one case and Poisson in the other. Let the probability level for measurement one be x_1 and the level for measurement two be x_2.

Consider the statistic $x = x_1 x_2$. What is the distribution of x, given that x_1 and x_2 are from flat distributions?

$$f(x)dx = \iint\limits_{x_1 x_2 = x = \text{constant}} (1 \times 1)\, dx_1\, dx_2. \tag{16.43}$$

The Jacobian of the transformation from $x_1 x_2$ to $x_1 x$ is easily seen to be $1/x_1$ and we obtain

$$f(x)\, dx = \left(\int_x^1 1 \times 1 \times 1/x_1\, dx_1 \right) \cdot dx = -\log(x)\, dx, \tag{16.44}$$

$$F(x_0) = \text{prob. } x < x_0 = \int_0^{x_0} f(x)\, dx = - \int_0^{x_0} \log(x)\, dx = x_0(1 - \log(x_0)). \tag{16.45}$$

$F(x_0)$ is the probability level of the product $x_1 x_2$, where x_1 and x_2 are the individual probability levels of the two tests. This method assumes that x_1 and x_2 are uncorrelated.

There are other limitations to the method. If the second measurement is not sensitive to any of the alternative hypotheses you are examining, then the combination only dilutes the effectiveness of the first probability level. Thus, in the example above, if the total number of events expected is the same for any of the possible hypotheses, then x_1 alone would be a better statistic to use than x. Each probability level should be examined separately in addition to examining the combined probability level. It sometimes is useful to examine, using computer Monte Carlo techniques, whether the combined statistic indeed adds information for the problem at hand.

The method is easily generalized to more than two tests. The density function for n tests $(x = x_1 x_2 \cdots x_n)$ is

$$f_n(x) = |\log|^{n-1}(x)/(n-1)!. \tag{16.46}$$

From this, it is easy to show by integration by parts and induction that the distribution function for n tests is

$$F_n(x) = x \sum_{j=0}^{n-1} \frac{1}{j!} |\log|^j (x). \qquad (16.47)$$

An ad hoc technique known as the elitist method has been introduced to help avoid the problem of dilution of results when some tests are better than others (Janot and Le Diberder 1997). Instead of the statistic $x_1 x_2 \cdots x_n$, where x_i is the probability level of test i, one uses $x_1^{a_1} x_2^{a_2} \cdots x_n^{a_n}$. The exponents are determined by Monte Carlo techniques to maximize the sensitivity. The resulting distribution of the product must then be obtained by numerical integration from the known individual distributions.

16.6 Exercises

16.1 Show that (16.36), is the solution for

$$\frac{\partial \chi^2(x)}{\partial x}\bigg|_{x=x_f} = 0$$

16.2 Show that (16.40), is the solution for (16.38) Hint: Add and subtract x. Then expand the terms and evaluate the cross term using (16.36).

References

Böck R (1960) Cern yellow report 60–30. Tech. rep., CERN, CH-1211, Geneva 23, Switzerland
Böck R (1961) Tech. rep., CERN, CH-1211, Geneva 23, Switzerland
Cowan G (2018) Statistics, in the review of particle physics. Phys Rev D 98(39):1–41
James F (2006) Statistical methods in experimental physics, 2nd Edn. World Scientific Publishing Co. Pte. Ltd, Singapore
Janot P, Le Diberder F (1997) Combining limits. Tech. rep., CERN, CH-1211, Geneva 23, Switzerland
Roe B (2013) Effect of correlations between model parameters and nuisance parameters when model parameters are fit to data
Ronne B (1968) Cern yellow report 64–13. Tech. rep., CERN, CH-1211, Geneva 23, Switzerland
Wilks S (1946) Mathematical statistics. Princeton University Press, Princeton, NJ

Chapter 17
Bartlett S Function; Estimating Likelihood Ratios Needed for an Experiment

Abstract We examine in this chapter some miscellaneous tools. If our distribution function is very far from a normal distribution, estimates can be biased, and also it is sometimes hard to interpret the non-gaussian errors that result. The Jacknife is a method to reduce the bias of an estimate. The use of the Bartlett S function is a technique to introduce new variables to make the distribution function closer to normal. The second topic treated in this chapter is of use when we are setting up an experiment. It is often crucial to estimate how long the experiment will take, how many events will be required to get the desired result. The second part of this chapter discusses how to estimate in advance of the experiment the number of events required.

17.1 The Jacknife

Sometimes we have an estimate that is biased. We first consider a method to reduce the bias in an estimate. It can be quite useful when applicable. It is called the "jacknife" and was proposed by Quenouille (1956). We follow the treatment by Kendall and Stuart (1979). Suppose we have a set of n trials and from this set we calculate an estimate p_n of a parameter whose true value is p. Further, suppose that the estimate is biased although asymptotically unbiased. We will suppose that the bias can be represented by power series in inverse powers of n

$$\mathbb{E}(p_n) - p = \sum_{s=1}^{\infty} a_s/n^s, \tag{17.1}$$

where the a_s are not functions of n although they may be functions of p.

Given this sample of n observations, we can make n samples each of $n - 1$ observations by omitting each observation in turn. For each subsample, we can determine our estimate. Suppose we do this and let $p_{n-1,\mathrm{AV}}$ be the average of these n estimates, each based on $n - 1$ trials. We now consider the new statistic

© Springer Nature Switzerland AG 2020
B. P. Roe, *Probability and Statistics in the Physical Sciences*,
Undergraduate Texts in Physics,
https://doi.org/10.1007/978-3-030-53694-7_17

$$p'_n = np_n - (n-1)p_{n-1,\text{AV}} = p_n + (n-1)(p_n - p_{n-1,\text{AV}}). \tag{17.2}$$

We can then easily calculate from the above equations that

$$\mathbb{E}(p'_n) - p = a_2 \left[\frac{1}{n} - \frac{1}{n-1}\right] + a_3 \left[\frac{1}{n^2} - \frac{1}{(n-1)^2}\right] + \cdots = -\frac{a_2}{n^2} - O(n^{-3}). \tag{17.3}$$

Thus, whereas p_n was biased to order $1/n$, p'_n is only biased to order $1/n^2$. This process can be carried further. Using definitions analogous to the above, we find that the parameter

$$p''_n = [n^2 p'_n - (n-1)^2 p'_{n-1,\text{AV}}]/[n^2 - (n-1)^2] \tag{17.4}$$

is only biased to order $1/n^3$.

This gives us a method of reducing the bias in our estimates of a parameter. It can be shown that if the variance of p_n is of order $1/n$, then the variance of p'_n will asymptotically be the same as that of p_n. This is not true for further order corrections. This method does not address the question of whether the estimate has an approximately normal distribution.

17.2   Making the Distribution Function of the Estimate Close To Normal; the Bartlett S Function

Consider what happens if we change variables in a maximum likelihood problem. Suppose we are measuring the decay of a moving particle (decay in flight). We could measure the lifetime τ or the decay length $= k/\tau$. Will we get the same maximum likelihood estimate? In general, we consider two representations of the parameters, α and μ, where $\alpha = \alpha(\mu)$. Then we want

$$\frac{\partial}{\partial \mu}(\log \mathcal{L}) = \frac{\partial}{\partial \alpha}(\log \mathcal{L}) \frac{d\alpha}{d\mu} = 0. \tag{17.5}$$

We see that $(\partial/\partial\mu)(\log \mathcal{L})$ will be zero when $(\partial/\partial\alpha)(\log \mathcal{L})$ is zero. Ignoring extra zeros or poles from $\partial\alpha/\partial\mu$, we have the same solutions: $\alpha^* = \alpha(\mu^*)$. If one uses a simple mean rather than a maximum likelihood solution, then the mean will vary with the choice of representation. The use of the median would be better. The maximum likelihood estimate is independent of the representation; unfortunately, the relation of the errors in the estimate is not that simple. For α, we have

$$\sqrt{\mathbb{E}\left\{\left(\frac{\partial \log f}{\partial \alpha}\right)^2_{\alpha=\alpha_0}\right\}} \sqrt{n}(\alpha^* - \alpha_0) \tag{17.6}$$

is asymptotically normal $(0,1)$. For μ, we have

$$\sqrt{\mathbb{E}\left\{\left(\frac{\partial\ \log\ f}{\partial\mu}\right)^2_{\mu=\mu_0}\right\}}\ \sqrt{n}(\mu^* - \mu_0) \qquad (17.7)$$

is asymptotically normal $(0,1)$. However,

$$\frac{\partial}{\partial\mu}(\log\ f) = \frac{\partial}{\partial\alpha}(\log\ f)\frac{d\alpha}{d\mu}. \qquad (17.8)$$

We note $d\alpha/d\mu$ is not a function of x and may be taken outside the expectation value. Equation(17.7) then becomes

$$\sqrt{\mathbb{E}\left\{\left(\frac{\partial\ \log\ f}{\partial\alpha}\right)^2_{\alpha=\alpha_0}\right\}}\ \sqrt{n}\left(\left(\frac{\partial\alpha}{\partial\mu}\right)_{\alpha=\alpha_0}(\mu^* - \mu_0)\right). \qquad (17.9)$$

Unless α is equal to μ times a constant, the two expressions for errors involving μ are different. Hence, we have a different variable to be taken as normal. Both expressions are normal $(0,1)$ asymptotically, but clearly, it is useful to make a choice so that the distribution at moderate n is as close to normal as possible. Thus, some care and thought in the choice of variables is very helpful.

Figure 17.1 shows the path of a charged particle through a constant magnetic field perpendicular to the path of the particle. The path is an arc of a circle.

If the sagitta s is such that $s << R$, then we have

Fig. 17.1 Definition of sagitta for a charged particle curving in a magnetic field

$$R^2 = (R - s)^2 + \frac{L^2}{4},$$

$$2Rs \cong \frac{L^2}{4},$$

$$s \cong \frac{L^2}{8R}.$$

However, we know that for this situation, R is proportional to the momentum p. Hence,

$$s = C/p,$$

where C is a constant. Suppose we determine the path of a particle by measuring its position at several points and suppose further that our measuring errors at each point are roughly normal. Then we see that it is much more likely that s, and hence $1/p$, will be normally distributed than p for moderate n. In fact, a measurement error can even give the wrong sign for p corresponding to a huge change (through infinity) for p, while only a small change (through 0) for s. In such situations, the most important thing to do is to choose as a variable the quantity most normally distributed.

M.S. Bartlett has suggested a slight modification of the maximum likelihood method, which can help for the above kind of problem in certain situations. This method was originally formulated for decaying particles, which have an exponential distribution of decay times. (See Exercise 17.2.) Suppose we find the estimate in the standard way, but to find errors, we consider

$$S(\alpha) = \frac{\partial w / \partial \alpha}{\sqrt{\mathbb{E}\left\{-\partial^2 w / \partial \alpha^2\right\}}}, \qquad (17.10)$$

where, again, $w = \log \mathcal{L}$. Note the order of events in the denominator. First, the derivative is taken and then the expectation value. We will show that S has mean 0 and variance 1. Unlike the previous expression, we use $\partial w / \partial \alpha$ instead of $\alpha^* - \alpha$ and we use a different square root expression. We will see directly that this has zero expectation value and unit variance. Bartlett suggests that, with a modification to be discussed below, this is a variable more closely normal for moderate n than the first expressions we wrote down in the chapter.

$$\mathbb{E}\left\{\frac{\partial w}{\partial \alpha}\right\} = \int \sum_{i=1}^{n} \frac{\partial \log f_i}{\partial \alpha} f_i \, dx = \sum_{i=1}^{n} \int \frac{1}{f_i} \frac{\partial f_i}{\partial \alpha} f_i \, dx$$

$$= \sum_{i=1}^{n} \int \frac{\partial f_i}{\partial \alpha} \, dx = \sum_{i=1}^{n} \frac{\partial}{\partial \alpha} \int f_i \, dx \qquad (17.11)$$

$$= \sum_{i=1}^{n} \frac{\partial}{\partial \alpha} 1 = 0.$$

Let us examine the variance.

$$\text{variance}\left(\frac{\partial w}{\partial \alpha}\right) = \int \left(\frac{\partial w}{\partial \alpha}\right)^2 \mathcal{L}\, dx_1\, dx_2 \cdots dx_n$$

$$= \sum_{i=1}^{n} \int \left(\frac{\partial \log f_i}{\partial \alpha}\right)^2 f_i\, dx_i. \tag{17.12}$$

This relation follows since all f_j, $j \neq i$, integrate to give one, and all cross terms vanish since we take the trials to be independent.

We note that if we specialize to the case where all trials are of the same form (all f_i are the same), then

$$\text{variance}\left(\frac{\partial w}{\partial \alpha}\right) = n\mathbb{E}\left\{\left(\frac{\partial \log f}{\partial \alpha}\right)^2\right\} = \frac{1}{\text{asymptotic variance}(\alpha^* - \alpha)}. \tag{17.13}$$

This shows that $S(\alpha)$ indeed is very similar to our initial equation in this chapter, but with $\alpha^* - \alpha$ replaced by $\partial w/\partial \alpha$. However, the present procedure is valid even if all of the f_i are not of the same form.

Consider

$$\mathbb{E}\left\{-\frac{\partial^2 w}{\partial \alpha^2}\right\} = \int -\frac{\partial^2 w}{\partial \alpha^2} \mathcal{L}\, dx_1\, dx_2 \cdots dx_n = -\sum_{i=1}^{n} \int \frac{\partial^2 \log f_i}{\partial \alpha^2} f_i\, dx_i$$

$$= -\frac{\partial}{\partial \alpha} \sum_{i=1}^{n} \int \frac{\partial \log f_i}{\partial \alpha} f_i\, dx_i + \sum_{i=1}^{n} \int \frac{\partial \log f_i}{\partial \alpha} \frac{\partial f_i}{\partial \alpha}\, dx_i.$$

The first term in this expression is just

$$-\frac{\partial}{\partial \alpha}\mathbb{E}\left\{\frac{\partial w}{\partial \alpha}\right\} = 0.$$

Also,

$$\frac{\partial f}{\partial \alpha} = \frac{1}{f}\frac{\partial f}{\partial \alpha} f = \frac{\partial \log f}{\partial \alpha} f.$$

Hence,

$$\mathbb{E}\left\{-\frac{\partial^2 w}{\partial \alpha^2}\right\} = \sum_{i=1}^{n} \int \left(\frac{\partial \log f_i}{\partial \alpha}\right)^2 f_i\, dx_i = \text{variance}\left(\frac{\partial w}{\partial \alpha}\right). \tag{17.14}$$

We have, therefore, shown S indeed has zero mean and unit variance. Bartlett (1953a; 1953b) suggests an improvement to this function to reduce skewness. We have made a slight improvement in his approximation. Define

$$\hat{S} = \frac{1}{b}[S - a(S^2 - 1)], \qquad (17.15)$$

$$a = \frac{\gamma_1}{3(\gamma_2 + 2)},$$

$$b = \sqrt{1 - \frac{5}{9}\frac{\gamma_1^2}{\gamma_2 + 2}},$$

where γ_1 and γ_2 are the coefficient of skewness and the kurtosis of $dw/d\alpha$ and, hence, of S.

It can easily be shown now that $\hat{S}$ has zero mean and unit variance. In the approximation that γ_1 is small, $\hat{S}$ has zero skewness. The distribution of $\hat{S}$ is a better approximation to a normal distribution for moderate n than the distribution of the expression in the initial equation in this chapter. Assuming $\hat{S}$ to be normal, we estimate confidence limits on α by examining the values of $\hat{S}$ to which they correspond.

For practical problems, we would like to calculate the coefficient of skewness in terms of the first power of derivatives of w. Consider the following identities:

$$\int \frac{\partial^3 w}{\partial \alpha^3} \mathcal{L}\, dx_1\, dx_2 \cdots dx_n$$

$$= \frac{\partial}{\partial \alpha}\left(\sum_{i=1}^{n} \int \frac{\partial^2 \log f_i}{\partial \alpha^2} f_i\, dx_i\right) - \sum_{i=1}^{n} \int \frac{\partial^2 \log f_i}{\partial \alpha^2}\frac{\partial \log f_i}{\partial \alpha} f_i\, dx_i.$$

$$\qquad (17.16)$$

$$2\int \frac{\partial^2 \log f_i}{\partial \alpha^2}\frac{\partial \log f_i}{\partial \alpha} f_i\, dx_i$$

$$= \frac{\partial}{\partial \alpha}\left(\int \left(\frac{\partial \log f_i}{\partial \alpha}\right)^2 f_i\, dx_i\right) - \int \left(\frac{\partial \log f_i}{\partial \alpha}\right)^3 f_i\, dx_i.$$

$$\qquad (17.17)$$

Note that the second term on the right in the first equation is the same as the term on the left in the second equation except for a factor of 2 and the sum over i.

We also use the result essentially proven above

$$\int \left(\frac{\partial \log f_i}{\partial \alpha}\right)^2 f_i\, dx_i = \int -\frac{\partial^2 \log f_i}{\partial \alpha^2} f_i\, dx_i. \qquad (17.18)$$

Then we can easily show

$$\sum_{i=1}^{n} \int \left(\frac{\partial \log f_i}{\partial \alpha}\right)^3 f_i\, dx_i = \mathbb{E}\left\{\left(\frac{\partial w}{\partial \alpha}\right)^3\right\} \equiv \mu_3; \qquad (17.19)$$

$$\mu_3 = 3\frac{\partial}{\partial\alpha}\mathbb{E}\left\{-\frac{\partial^2 w}{\partial\alpha^2}\right\} + 2\mathbb{E}\left\{\frac{\partial^3 w}{\partial\alpha^3}\right\}. \tag{17.20}$$

We used the fact that the cross terms vanish in obtaining the above relation. We, thus can calculate μ_3 in terms of the first power of derivatives of w. The coefficient of skewness is given by

$$\gamma_1 = \frac{\mu_3}{(\mu_2)^{3/2}} = \frac{\mu_3}{\left(\mathbb{E}\left\{-\partial^2 w/\partial\alpha^2\right\}\right)^{3/2}}. \tag{17.21}$$

17.3 Estimating in Advance the Number of Events Needed for an Experiment

Suppose we wish to plan an experiment with independent events to test between two hypotheses, A and B. How many events will we be likely to need? Assume we wish a given likelihood ratio. We then estimate the number of events, n, from

$$R = \mathcal{L}_A/\mathcal{L}_B, \tag{17.22}$$

$$\mathbb{E}\{\log R\} \text{ if } A \text{ is true} = n\mathbb{E}\left\{\log\frac{f_A}{f_B}\right\} = n\int\left(\log\frac{f_A}{f_B}\right)f_A\,dx. \tag{17.23}$$

Here $f_A(f_B)$ is the density function if hypothesis $A(B)$ is true. Next, we examine the variance of $\log R$.

$$\log R = \sum_{i=1}^{n}\log\left(\frac{f_A}{f_B}\right)_i.$$

Let

$$y_i = \left(\log\frac{f_A}{f_B}\right)_i.$$

Then

$$y = \log R = \sum_{i=1}^{n}y_i,$$

$$\text{variance}(y) = \mathbb{E}\left\{\left(\sum_{i=1}^{n}y_i - \sum_{k=1}^{n}\mathbb{E}\{y_k\}\right)^2\right\}$$

$$= \mathbb{E}\left\{\sum_{i=1}^{n}(y_i - \mathbb{E}\{y_i\})^2 + \sum_{\substack{i,j \\ i\neq j}}(y_i - \mathbb{E}\{y_i\})(y_j - \mathbb{E}\{y_j\})\right\}.$$

The second term is zero since we have independent trials.

$$\text{variance}(y) = \sum_{i=1}^{n}(y_i - \mathbb{E}\{y_i\})^2 = \sum_{i=1}^{n} \text{variance}(y_i). \qquad (17.24)$$

This is a generally useful relation for independent events (we have already noted this point in Chap. 3 when we discussed multiple scattering). Here we have

variance($\log R$) if A is true

$$= n\left(\int\left(\log \frac{f_A}{f_B}\right)^2 f_A\, dx - \left[\int\left(\log \frac{f_A}{f_B}\right)f_A\, dx\right]^2\right). \qquad (17.25)$$

As an example, suppose we wish to determine whether a certain angular distribution is flat or proportional to $1 + x$, where $x = \cos\theta$.

$$\text{For Case } A, \qquad f_A = \frac{1}{2}.$$

$$\text{For Case } B, \qquad f_B = \frac{1+x}{2}.$$

Suppose first that hypothesis A is true.

$$\mathbb{E}\{\log R\}_A = n\mathbb{E}\left\{\log \frac{f_A}{f_B}\right\}$$

$$= n\int_{-1}^{1}\left(\log \frac{1}{1+x}\right)\frac{1}{2}\, dx;$$

$$\mathbb{E}\{\log R\}_A = n(1 - \log 2) = 0.306n. \qquad (17.26)$$

For an individual term in $\log R$

$$\mathbb{E}\left\{\left(\log \frac{f_A}{f_B}\right)^2\right\}_A = \int_{-1}^{1}[\log(1 + x)]^2\frac{1}{2}\, dx = (\log 2)^2 - 2\log 2 + 2 \cong 1.09.$$

Using Eq. 17.25, we then have

$$\text{variance}(\log R)_A = n[1.09 - (0.306)^2] \cong n. \qquad (17.27)$$

Next, suppose that hypothesis B is true

$$\mathbb{E}\{\log R\}_B = n\int_{-1}^{1}\left(\log \frac{1}{1+x}\right)\frac{1+x}{2}\, dx; \qquad (17.28)$$

$$\mathbb{E}\{\log R\}_B = n\left(\frac{1}{2} - \log 2\right) \cong -0.194n.$$ (17.29)

For an individual term in $\log R$

$$\mathbb{E}\left\{\left(\log \frac{f_A}{f_B}\right)^2\right\}_B = \int_{-1}^{1} [\log(1+x)]^2 \frac{1+x}{2} dx = (\log 2)^2 - \log 2 + \frac{1}{2} \cong 0.29.$$

$$\text{variance}(\log R)_B = n[0.29 - (0.194)^2] \cong 0.25n.$$ (17.30)

Thus, we have the following table:

A Correct	B Correct
$\mathbb{E}\{\log R\} = 0.306n$	$-0194n$
variance $= n$	$0.25n$

Suppose we ask to have a mean $\log R$ corresponding to a likelihood ratio of 10^5 if A is true. Then $\log 10^5 = 11.5$ and, using the above equations, we need 38 events. If $\log R$ were one standard deviation low, it would be $11.5 - \sqrt{38} = 5.4$, corresponding to a likelihood ratio of about 220. Next, suppose B were true and we had 38 events. Then we obtain $\mathbb{E}\{\log R\} = -7.37$ or $R \cong 1/1600$. If B were true and we were one standard deviation high, we would get $-7.37 + \sqrt{0.25 \times 38} = -4.29$ or $R \cong 1/73$.

Note that, for small n, the distribution function for $\log R$ is not normal but has longer tails. For large n, the central limit theorem applies and the distribution of $\log R$ approaches a normal distribution. For 38 events, it should be fairly close to normal.

So far, it would appear 38 events is sufficient. It is too small a number if we require the right answer for a three standard deviation error. Suppose A were true and $\log R$ were three standard deviations low. We would then get $11.5 - 3\sqrt{38} = -7$. This result actually favors hypothesis B and, hence, this would be a dangerous test. We should ask for more data. We note, however, that we could be very lucky. If A were true and we got a few events very near $x = -1$, then even with a small number of events, we would have excellent discrimination.

We have examined some further statistics tools in this chapter. The use of the jacknife or the Bartlett S function are techniques for changing variables to make the distribution function of a fitted variable closer to the normal distribution as an aid in interpreting errors. We discussed likelihood ratios as a method of estimating in advance of an experiment the number of events that would be required to obtain a desired level of distinction between alternate hypotheses.

17.4   Exercises

17.1 Suppose we wish to show that a coin is reasonably fair. To this end, we will try to distinguish between the hypotheses

(a) The probability of a toss of the coin yielding heads is 0.5.
(b) The probability of a toss of the coin yielding heads differs from 0.5 by at least 0.1.

Estimate the number of trials that will be needed to distinguish between these two hypotheses at the 95% level.

17.2 Find the Bartlett S function for the case that one makes n measurements of the time of an event on a distribution whose frequency function varies with time as $e^{-t/\tau}$ from $t = 0$ to $t = T$ and is 0 for $t > T$. Only estimate τ; take T as a fixed parameter.

17.3 We make n measurements of x from a normal distribution and estimate the variance using the biased estimate, variance $= \sum (x_i - x_{AV})^2/n$, where $x_{AV} = \sum x_i/n$. For $n = 5$ and $n = 10$, generate events using Monte Carlo methods on a computer and calculate the first and second order jacknife estimates of the variance.

References

Bartlett M (1953a) Approximate confidence intervals. Biometrika 40:12
Bartlett M (1953b) On the statistical estimation of mean life-times. Philos Mag 44:249
Kendall SM, Stuart A (1979) The advanced theory of statistics, vol 2, inference and relationship, 4th edn. Griffin C. and Company, London and High Wycombe
Quenouille M (1956) Notes on bias in estimation. Biometrika 43:353

Chapter 18
Interpolating Functions and Unfolding Problems

Abstract Often we have a set of n data points (z_i, y_i), $i = 1, \ldots, n$, and want to find a function that matches the measured points and smoothly interpolates between them. For example, we might have measured the distortions on a lens at a series of different angles away from the axis and wish to find a smooth curve to represent the distortions as a function of angle. In this chapter, we will discuss this problem especially concentrating on spline functions. The other topic discussed in this chapter is unfolding data. Data, as obtained, is often smeared by statistical and systematic errors. Usually, it is best to fit using the smeared data. However, sometimes it is useful to try to remove the smearing. Several methods for unfolding the data are described.

18.1 Interpolating Functions

One method for smoothly interpolating between and set of measured points is to use an expansion of basis functions. This method can work quite well if we fit using the regularized χ^2 method described in Sect. 15.3.

One choice of basis functions is the use of spline functions and we will discuss these in the next sections

18.2 Spline Functions

A spline function, $y = S(z)$, of kth degree is a $k - 1$ differentiable function coinciding on every subinterval (z_i, z_{i+1}) with a polynomial of degree k (Blobel 1984). The set of abscissa points, z_i, are called "knots". At every knot, the definition requires that the values of the polynomials of the adjacent intervals and their first $k - 1$ derivatives must match.

Because of the endpoint effects, the coefficients of the polynomials are not completely fixed. At the inner knots, there are $(n - 2) \times k$ conditions from matching the polynomial and derivative values at the knots. There are n further conditions for matching the polynomials to the values y_i at the knots. Since a polynomial of kth

© Springer Nature Switzerland AG 2020
B. P. Roe, *Probability and Statistics in the Physical Sciences*,
Undergraduate Texts in Physics,
https://doi.org/10.1007/978-3-030-53694-7_18

degree has $k + 1$ coefficients, we have $(n - 2)k + n$ conditions and $(n - 1)(k + 1)$ coefficients to determine. We lack $k - 1$ conditions.

Cubic splines are among the most commonly used splines and we will concentrate on them for now. For cubic splines, we lack two conditions. There are several standard choices to determine them

(a) $S''(z_1) = 0$; $S''(z_n) = 0$. Natural spline.
(b) $S'(z_1) = y_1'$; $S'(z_n) = y_n'$. Complete spline. (This choice requires known values for the initial and final first derivatives.)
(c) $S''(z_1) = y_1''$; $S''(z_n) = y_n''$. (This choice requires known values for the initial and final second derivatives.)
(d) $S'''(z)$ continuous across z_2 and z_{n-1}. Not-a-knot condition.

The last condition is called the not-a-knot condition because it forces the polynomials for the first two regions and for the last two regions to each be identical. Thus, there is no change in polynomial across the second or the penultimate knot.

Recall that we have used

$$\int_a^b [G''(z)]^2 \, dz$$

as a measure of the curvature of a function in Sect. 15.4. Using this measure, it can be shown that the natural spline gives the function interpolating the z_i, y_i with the least possible curvature. (This same function approximately minimizes the strain energy in a curved elastic bar,

$$\int_a^b \frac{[G''(z)]^2}{[1 + G'(z)]^{5/2}} \, dz. \tag{18.1}$$

This is the origin of the name "spline function".)

Smoothness, i.e., lack of curvature, is not the sole criterion, however. If the derivatives are known at the beginning and at the end, conditions (b) or (c) can lead to improved accuracy. If the points, z_i, y_i are points on a smooth function $y = G(z)$, then, for the not-a-knot choice, it can be shown that for the pth derivative of the function,

$$|G^p(z) - S^p(z)| < B, \quad B \propto (\text{knot spacing})^{4-p}, \quad p = 0, 1, 2, 3. \tag{18.2}$$

The natural spline condition produces an error proportional to (knot spacing)^2 near the ends of the interval. This goes to zero more slowly than the (knot spacing)^4 of the first term above ($p = 0$) as the knot spacing decreases. The not-a-knot condition avoids this problem and is recommended as a general condition.

18.3 B-Splines

The basis spline, or B-spline representation, provides a different, but equivalent, representation for the spline functions. Each B-spline is non-zero only in a limited region and the function S is the sum of the B-splines used.

$$S(z) = \sum_j a_j B_{j,k}(z), \qquad (18.3)$$

where the B-splines, $B_{j,k}$, are of order k. The B-splines are defined by a recurrence relation. They are defined over a non-decreasing sequence, t_j, of knots. For $k = 1$,

$$B_{j,1} = \begin{cases} 1 & t_j \leq z \leq t_{j+1}, \\ 0 & \text{otherwise}. \end{cases} \qquad (18.4)$$

The recurrence relation consists of positive linear combinations of positive quantities for $k > 1$

$$B_{j,k} = \frac{z - t_j}{t_{j+k-1} - t_j} B_{j,k-1}(z) + \frac{t_{j+k} - z}{t_{j+k} - t_{j+1}} B_{j+1,k-1}(z). \qquad (18.5)$$

B-splines can be non-zero only in the region $t_j < z < t_{j+k}$. Thus, in any one interval, only k B-splines can be non-zero. At any particular z value, the sum of the B-splines is 1. If we use B-splines to fit data rather than just for interpolation, we note that using this parametrization, linear least squares fits to data can be made.

Again we will specialize to cubic B-splines, here given by $k = 4$. For equidistant knots, we denote $b_j(z) \equiv B_{j,4}(z)$. The use of n subintervals requires $m = n + 3$ B-splines, b_j, and $m + 1$ knots with $t_4 = a$; $t_{m+1} = b$. If d is the distance between knots $[d = (b - a)/(m - 3)]$, b_j is given by

$$
\begin{aligned}
b_j(z) &= \eta^3/6 & \eta &= (z - t_j)/d, \\
& & t_j &\leq z \leq t_{j+1}; \\
b_j(z) &= [1 + 3(1 + \eta(1 - \eta))\eta]/6 & \eta &= (z - t_{j+1})/d, \\
& & t_{j+1} &\leq z \leq t_{j+2}; \\
b_j(z) &= [1 + 3(1 + \eta(1 - \eta))(1 - \eta)]/6 & \eta &= (z - t_{j+2})/d, & (18.6) \\
& & t_{j+2} &\leq z \leq t_{j+3}; \\
b_j(z) &= (1 - \eta)^3/6 & \eta &= (z - t_{j+3})/d, \\
& & t_{j+3} &\leq z \leq t_{j+4}; \\
b_j(z) &= 0, & & \text{Otherwise}.
\end{aligned}
$$

Next, we need to determine the m coefficients a_j of the cubic B-splines. $m - 2$ equations are obtained from the interpolation conditions:

$$6y_j = a_{j-3} + 4a_{j-2} + a_{j-1}, \quad j = 4, \ldots, m+1. \tag{18.7}$$

Using the not-a-knot condition yields the last two needed relations.

$$-a_1 + 4a_2 - 6a_3 + 4a_4 - a_5 = 0, \tag{18.8}$$

$$-a_{m-4} + 4a_{m-3} - 6a_{m-2} + 4a_{m-1} - a_m = 0. \tag{18.9}$$

18.4   Unfolding Data

We often have the problem of unfolding results. For example, we may be collecting events and measuring a quantity, z, for each event. The apparatus often is biased and has an efficiency that is a function of z. Furthermore, there may be a finite resolution and the measured value, z', may not correspond to the true value z. Finally, there may be a background of accidental events, $B(z')$. If z goes from a to b, we have for the expected density function

$$y(z') = \int_a^b A(z', z) f(z) \, dz + B(z'), \tag{18.10}$$

where A is the combined resolution and bias function, i.e., the probability that an event with real parameter z will appear to have parameter z'.

Usually, it is best to fit this smeared version of the theory to the data. This will, in most instances, give the best and most statistically accurate value of the parameters being sought.

However, after making a fit, it is sometimes useful to try to visualize the unfolded distribution, i.e. what the distribution would look like if there were no smearing. Suppose we have measured $y(z')$ and wish to deduce the true density function $f(z)$. We assume that the resolution-bias function $A(z', z)$ and the background function $B(z')$ are known or can be measured independently. Imagine the data to be in the form of a histogram with k bins.

One simple method for unfolding resolution starts with an expected distribution and the known resolution-bias function. The probability of going from one histogram bin to another is then calculated. Let β_{ij} be the probability that an event which is really in bin i is measured to be in bin j. If the true number of events in bin i is n_i^T, and the number measured in bin i is n_i^M, then the expectation values of the number of events measured in bin j is $\mathbb{E}(n_i^M) = \sum_{i=1}^k \beta_{ij} n_i^T$. If the variation in the expected number of events between adjacent bins is not too large, i.e, if the bin width is small, then β_{ij} will be largely independent of the assumed distribution. We can then invert the matrix β. Replacing expectation values by the actual measured values, we obtain, $n_i^U = \sum_{j=1}^k (\beta^{-1})_{ij} n_i^M$, where n_i^U is the unfolded value for bin i and is an approximation for the true value. It can be shown that this estimate is the most efficient unbiased estimate possible. Nonetheless, this is not a favored method because it has very large fluctuations, reminiscent of the Gibbs phenomenon.

Since this is the most efficient unbiased estimate possible, we must resort to biased estimates to do better. The solution is a regularization technique similar to that introduced to mitigate the Gibbs phenomenon. We minimize not χ^2, but $\chi^2_{\text{reg}} = \chi^2 + \tau r$, where τ is a constant and r is a function chosen to emphasize smoothness. τ is chosen to minimize some combination of bias and statistical error. Several choices for r are in common use as are several criteria for choosing τ. We have discussed some of these choices in Sects. 15.3, 15.4. Here we will use the Tikhonov regularization scheme discussed in Sect. 15.3. It corresponds to minimizing a measure of the average square of the local curvature.

We start by expanding $f(z)$ using a set of basis functions $\phi_j(z)$ (Blobel 1984).

$$f(z) = \sum_{j=1}^{m} \alpha_j \phi_j(z). \tag{18.11}$$

We can now define a smeared set of basis functions, ϕ_j^s.

$$\phi_j^s(z') = \int_a^b A(z', z) \phi_j(z) \, dz. \tag{18.12}$$

Then we have

$$y(z') = \int_a^b A(z', z) f(z) \, dz + B(z') = \sum_{j=1}^{m} \alpha_j \phi_j^s(z') + B(z'). \tag{18.13}$$

Often we have to determine $A(z', z)$ by a Monte Carlo method, generating events at values z and then smearing them to obtain a z'. To obtain ϕ_j^s above, we give each event a weight proportional to $\phi_j(z)$. We here assume that the $\phi_j(z) \geq 0$. If we also arrange that $\sum_{j=1}^{m} \phi_j(z) \equiv 1$, this corresponds to asking that the sum of the weights for each Monte Carlo event be 1.

We next divide the result into a histogram of n bins, integrating each term from z'_{i-1} to z'_i for the ith bin. Generally, we take $n > m$ and ϕ_j^s becomes ϕ_{jv}^s, an $m \times n$ matrix

$$y_v = \sum_{j=1}^{m} \alpha_j \phi_{jv}^s + B_v. \tag{18.14}$$

The general procedure is to now make a fit for the m parameters, α_j. The use of these parameters with the unsmeared ϕ_{jv} estimate the unsmeared distribution, the unfolded data. We use the regularization method described previously to make the fit. If we have a set of unweighted events, the variance of y_v is approximately y_v and the weight of the bin is $\omega_v \approx 1/y_v$

$$\chi^2 = \sum_{\nu=1}^{n} \left(y_{\nu \text{ meas}} - \sum_{j=1}^{m} \alpha_j \phi_{j\nu}^s - B_\nu \right)^2 \omega_\nu. \tag{18.15}$$

A convenient choice of basis functions $\phi_j(z)$ is often the cubic B-splines discussed above. The b_j given in Eq. 18.6, correspond to the ϕ_j and it is these b_j which are smeared according to the resolution function. The B-splines are already smoothed fits to data and only involve second order polynomials, not an infinite series. For this reason, it is often not necessary to add a regularization term. However, suppose for a particular problem it is necessary to add this term and that there are $m - 1$ cubic B-splines used for the fit. For this choice, the $tr(\alpha)$ term added to χ^2, indeed, can be represented as a quadratic form, $r(\alpha) = \alpha^T C \alpha$, where C is an $m - 1 \times m - 1$ matrix whose elements are

$$C_{11} = 2, C_{22} = 8, C_{33} = 14, \tag{18.16}$$

$$C_{12} = C_{21} = -3, C_{23} = C_{32} = -6, \tag{18.17}$$

$$C_{k,k+3} = C_{k+3,k} = 1, k \geq 1, \tag{18.18}$$

$$C_{k,k+1} = C_{k+1,k} = -9, k \geq 3, \tag{18.19}$$

$$C_{kk} = 16, k \geq 4, \tag{18.20}$$

$$C_{m-1,m-1} = 2C_{m-2,m-2} = 8, C_{m-3,m-3} = 14, \tag{18.21}$$

$$C_{m-2,m-1} = C_{m-1,m-2} = -3, C_{m-2,m-3} = C_{m-3,m-2} = -6. \tag{18.22}$$

$$\text{All other elements} = 0. \tag{18.23}$$

If the histogram involves an experimental smearing from bin-to-bin, then it is necessary to transfer to a set of variables in which the error matrix is diagonal. Sections 15.3 and 15.4 discuss a method of finding the solution to the resulting χ^2 function including, if necessary, the regulator term.

Sometimes it is desirable to express the result as pseudo-data points. The transformed basis function ϕ'_{m_0+1} will usually have m_0 zeros when using the regularization method (see Eq. 15.45). A convenient choice of $m - 1$ data points is then to integrate over the intervals between the maxima of ϕ'_{m_0+1}. This gives about as many points as have been found to be significant and has the effect of suppressing the contribution of the next term in the series. Each average value is a linear function of the terms α_j, making error propagation straightforward.

18.5 Further Suggestions for Unfolding

The first method was used with some success in the MiniBooNE experiment: Laird et al. (2007), Linen (2007), but real proofs are not available.

- D_i be the data, $i = 1, N$
- T_i be the theory
- S_{ij} be the smearing function, i.e., $< D_i > = S_{ij}T_j$
- U_i be the estimated "unfolded" data

Let $F_{\ell,j}$ be the fraction of data bin j to assign to the unfolded bin ℓ. The expected total number of events in data bin j given the theory T_i is

$$\sum_k S_{jk}T_k \equiv M_j; \quad F_{\ell,j} = S_{j\ell}T_\ell/M_j.$$

$$U_\ell = \sum_j F_{\ell,j}D_j$$

This can be used to estimate a new theory distribution and iterate.

There are errors due to data statistics, due to uncertainty in the smearing function, and uncertainty in the theory. We consider only the first iteration.

18.5.1 Uncertainty in the Data Statistics

$$\Delta U_\ell = \sum_j F_{\ell j}\Delta D_j = \sum_j F_{\ell j}\Delta < M_j > .$$

The covariance

$$c^U(statistics)_{mn} = < \Delta U_m \Delta U_n > = \sum_j F_{mj}F_{nj}M_j,$$

recognizing that for statistics, the different bins are independent. In the variance, the term $F_{mj}F_{nj}$ reduces the error as it averages fluctuations over several bins. This is quite unlike the unfolding technique by matrix inversion, which magnifies errors. However, as in any unfolding technique, there are strong bin-to-bin correlations.

18.5.2   Uncertainty in the Smearing Function

Assume that there are theory bin-to-bin correlations in the smearing functions. Let $\Delta S_{j\ell}$ be the uncertainty in the smearing function of theory bin ℓ into data bin j.

$$U_\ell = \sum_j F_{\ell j} D_j; \quad \Delta U_\ell = \sum_j \left(\frac{\Delta S_{j\ell} T_\ell D_j}{M_j} - \frac{S_{j\ell} T_\ell D_j \Delta M_j}{M_j^2} \right).$$

Note that $< D_j > = M_j$ for no theory error. Ignore the statistics and theory variations for the smearing calculation, so $D_j = M_j$. Further note that $\sum_j \Delta S_{j\ell} T_\ell = 0$, i.e., for a given theory bin the events go somewhere, so the sum of ΔS over all data bins is zero. Define

$$c^S_{jk;\ell m} = < \Delta S_{j\ell} \Delta S_{km} > \quad \text{and}$$

$$c^{S,M}_{jk} = < \Delta M_j \Delta M_k > = < \left(\sum_r \Delta S_{jr} T_r \right) \left(\sum_s \Delta S_{ks} T_s \right) > .$$

Then the cross terms vanish because they involve $\sum_j \Delta S_{j\ell} T_\ell$. The result is

$$c^U_{\ell m}(smearing) = \sum_{jk} \left(T_\ell T_m c^S_{jk;\ell m} + \frac{S_{j\ell} S_{km} T_\ell T_m}{M_j M_k} c^{S,M}_{jk} \right).$$

18.5.3   Uncertainty in the Theory

$$M_j = \sum_k S_{jk} T_k; \quad \Delta M_j = \sum_\ell S_{j\ell} \Delta T_\ell; \quad c^{M,T}_{jk} = \sum_{r,s} S_{jr} S_{ks} c^T_{rs},$$

where $c^T_{k\ell}$ is the covariance of the theory estimate.

$$U_\ell = \sum_j \frac{S_{j\ell} T_\ell D_j}{M_j}; \quad \Delta U_\ell = \sum_j \left(\frac{D_j S_{j\ell} \Delta T_\ell}{M_j} - \frac{D_j S_{j\ell} T_\ell \Delta M_j}{M_j^2} \right).$$

Ignore the statistics and smearing variations for the theory calculation so $D_j = M_j$.

$$c^U_{\ell m}(theory) = \sum_{jk}\left(S_{j\ell}S_{km}c^T_{\ell m} + \frac{S_{j\ell}S_{km}T_\ell T_m \sum_{r,s}S_{jr}S_{ks}c^T_{rs}}{M_j M_k}\right)$$

$$-S_{j\ell}S_{km}\left(\frac{T_m \sum_r S_{kr}c^T_{\ell r}}{M_k} + \frac{T_\ell \sum_r S_{jr}c^T_{mr}}{M_j}\right).$$

The total covariance of U is

$$c^U_{\ell m} = c^U_{\ell m}(statistics) + c^U_{\ell m}(smearing) + c^U_{\ell m}(theory),$$

where
$c^{S,\ell}_{j,k} =$ covariance for bins j, k of the smearing due to theory bin ℓ.

$$c^S_{jk;\ell m} = < \Delta S_{j\ell}\Delta S_{km} > \text{ and}$$

$$c^{S,M}_{jk} = < \Delta M_j \Delta M_k > = < \left(\sum_r \Delta S_{jr}T_r\right)\left(\sum_s \Delta S_{ks}T_s\right) > .$$

c^T_{mr} is the covariance between theory bins m and r due to theory uncertainty.

18.5.4 Another Suggested Unfolding Technique

This technique was suggested by a Statistics Department colleague, Woodroofe (2007). Replace S in the inversion process by a regularized version. One simple way to do this is to insert a ridge. In matrix notation, D and T and U are vectors and S is a matrix. S^T is the transpose of S. Insert a ridge. Let

$$U = T^0 + [S^T S - \lambda I_N]^{-1}S^T(D - ST^0),$$

where T^0 is a prior guess for T, I_N is the identity matrix, and λ is a smoothing parameter. An alternative is to replace I_N by the implicit prior guess for the covariance of D, which can be computed from T^0. The T^0 has to be specified; the λ can be specified or estimated. This is a lot like specifying a prior distribution. Observe that when $\lambda = 0$, $U = S^{-1}D$ and when $\lambda \to \infty$, $U \to T^0$.

If λ is specified, then the mean, bias and covariance matrix of U are

$$< U > = T^0 - [S^T S + \lambda I_N]^{-1}(S^T S)(T - T^0),$$

$$bias = \lambda[S^T S + \lambda I_N]^{-1}(T - T^0),$$

and

$$covariance\ matrix = [S^T S + \lambda I_N]^{-1}\Sigma[S^T S + \lambda I_N]^{-1},$$

where Σ is the covariance matrix of D.

Several other unfolding suggestions (and warnings) are given in references: Balasubramanian et al. (2019), Gaponenko (2019), Cousins et al. (2016), Andreeassen et al. (2019).

In this chapter, we have discussed problems of interpolating between data points and problems of unfolding experimental resolution from measured data. We have examined the use of cubic spline functions as one method to treat both of these problems.

18.6   Exercises

18.1 Consider the function $f(z) = \sin(z)$. Divide this into eight equal intervals, (nine knots) from $z = 0$ to $z = \pi$. Use the cubic B-spline method to interpolate between the nine values of z and compare the interpolated B-spline function with the actual value at the center of each interval. This exercise is best done on a computer.

18.2 Generate a Monte Carlo distribution of 1000 events where one-fourth of the events are background distributed according to $1.0 + 0.3E$ for $0 \leq E \leq 2$, one-half of the events have a Breit–Wigner distribution with $M = 1$, $\Gamma = 0.2$, and the remaining one-fourth have a Breit–Wigner distribution with $M = 1.5$, $\Gamma = 0.15$. Further suppose that there is an experimental resolution function which smears the Breit–Wigner parts of the distribution. The resolution function is a normal distribution with $\sigma = 0.10$. Generate and plot this distribution both with and without the effect of folding in this resolution function. Display the results as 50-bin histograms.

References

Andreeassen A, Komiske P et al (2019) A method to simultaneously unfold all observables
Balasubramanian R, Brenner L et al (2019) Statistical method and comparison of different unfolding techniques using roofit
Blobel V (1984) Unfolding methods in high-energy physics experiments. Technical report, DESY, D2000, Hamburg-52, Germany
Cousins RD, May SJ, Sun Y (2016) Should unfolded histograms be used to test hypotheses?
Gaponenko A (2019) A practical way to regularize unfolding of sharply varying spectra with low data statistics. Technical Report FERMILAB-PUB-19-262-PPD, Fermilab PO Box 500, Batavia IL, 60510-5011
Laird E, Patterson R, Tanaka H (2007) Miniboone Technical Note, tn222
Linen K (2007) Private communication
Woodroofe MB (2007) Private communication

Chapter 19
Beyond Maximum Likelihood and Least Squares; Robust Methods

Abstract The methods discussed previously are very powerful if the underlying distributions are close to normal. If the distributions of errors of the points on a curve, for example, are not normal and have long tails, then, as we noted in Chap. 17, estimates of goodness of fit may be seriously biased. The tests discussed in this chapter tend to be robust, with results that are independent of the particular distribution being tested. Furthermore, the least squares fit ignores bin-to-bin correlations. The Smirnov–Cramèr–Von Mises fit and the Kolmogorov–Smirnov fits look at the distribution function, not the density function. A number of other tests are also described.

19.1 Problems with a Least Squares Fit

The least squares method does not always extract all of the relevant information from data. An example of this using a histogram is given in Fig. 19.1.

A least squares fit will give the same χ^2 for both distributions assuming the hypothesis of constant f. However, we intuitively feel that in histogram A it is more likely that the constant f hypothesis is wrong. It might be argued that if the experimental χ^2 is not too bad, then it is hard to rule out that the arrangement of histogram A is accidental. Thus, if it were fit to a sloping line, the error on the slope from a χ^2 analysis might be large enough to intersect zero slope. However, it is precisely in these borderline problems that we would like to squeeze out as much information as possible. Several methods have been developed to attempt to go further by trying to get information on the correlation of deviations from bin-to-bin which the least squares method ignores.

19.2 Tests on the Distribution Function

These tests use as their starting points the distribution function (F), sometimes called the cumulative distribution function. If we plot the distribution function for the histograms of Fig. 19.1, we have the results shown in Fig. 19.2. The line shown is the

© Springer Nature Switzerland AG 2020
B. P. Roe, *Probability and Statistics in the Physical Sciences*,
Undergraduate Texts in Physics,
https://doi.org/10.1007/978-3-030-53694-7_19

Fig. 19.1 Two histograms having the same fit to the hypothesis of constant value for f when using the least squares method

Fig. 19.2 Distribution functions for data and for the fit to the histograms of Fig. 19.1

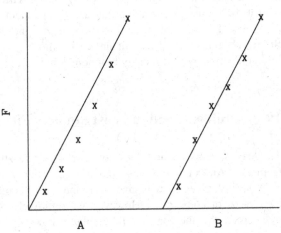

least squares fit for the constant frequency function. It is clear that we have a better chance of distinguishing A and B from this plot than from the preceding one.

The first test we discuss is known as the Smirnov–Cramèr–Von Mises goodness of fit test. Let

$$\omega^2 = \int_{-\infty}^{\infty} [F^*(x) - F(x)]^2 \, d'K(x). \tag{19.1}$$

$F^*(x)$ is the sample distribution function. $F(x)$ is the hypothesized distribution function. K can, in principle, be arbitrary, but we will take $K = F$ for our analysis. ω^2 is a measure of the absolute value of the deviation.

Assume the sample values are arranged in increasing order. We can then show (Fisz 1967) that for continuous $F(x)$, i.e., no binning

$$\omega^2 = \frac{1}{12n^2} + \frac{1}{n} \sum_{v=1}^{n} \left(F(x_v) - \frac{2v-1}{2n} \right)^2, \tag{19.2}$$

where n is the number of samples. The proof is left to the exercises. Next, we note that $F^*(x)$ is the sample probability function in n trials of an event of probability $F(x)$. Thus

$$\mathbb{E}\{(F^* - F)^2\} = \frac{F(1-F)}{n}. \tag{19.3}$$

This is not an expectation with respect to x. Here x is fixed and each of the n trials is a success if its result is less than x and a failure if its result is greater than x. From this, it can be shown (Kiefer 1959; Anderson and Darling 1952; Akerlof 1991) that

$$\mathbb{E}\{\omega^2\} = \frac{1}{6n},$$

$$\text{variance}(\omega^2) = \frac{4n-3}{180n^3}. \tag{19.4}$$

Note that this test is less dependent on small independent uncertainties x_v of individual samples than are the χ^2 tests.

The distribution of $n\omega^2$ is a non-normal distribution even in the limit of large n. It is, however, independent of F. The probability that $n\omega^2$ is greater than the observed value is known as the significance level of the test. Tables 19.1 and 19.2 list the significance levels for various $n\omega^2$ values (Kiefer 1959; Akerlof 1991).

A discussion of many related tests is given by Fisz (1967). We will only list (without proofs) a few of these.

Let D_n be the maximum absolute value of $F(x) - F^*(x)$ for all x given a sample of n. Suppose we are testing a hypothesis and have not estimated any parameters. Then

$$\lim_{n \to \infty} P\left(D_n < \frac{\lambda}{\sqrt{n}} \right) = Q(\lambda),$$

$$Q(\lambda) = \sum_{k=-\infty}^{\infty} (-1)^k e^{-2k^2\lambda^2}. \tag{19.5}$$

This relation is good for n greater than about 80. This is known as the Kolmogorov–Smirnov test. $Q(\lambda)$ is known as the Kolmogorov–Smirnov distribution. $Q(\lambda)$ is a monotonically increasing function. $Q(0) = 0$; $Q(\infty) = 1$. Tables 19.3 and 19.4 list some values of $Q(\lambda)$.

Suppose we wish to test whether two samples are from the same distribution. Let D_{n_1, n_2} be the maximum absolute value of $F_{n_1}^* - F_{n_2}^*$ for all x, where $F_{n_1}^*$ and $F_{n_2}^*$ are the sample distribution functions for the two independent samples

Table 19.1 Smirnov–Cramèr–Von Mises Test

$n\omega^2$	Signif	$n\omega^2$	Signif	$n\omega^2$	Signif	$n\omega^2$	Signif
0.00	1.000000	0.36	0.092383	0.72	0.011393	1.08	0.001600
0.01	0.999994	0.37	0.086832	0.73	0.010776	1.09	0.001516
0.02	0.996999	0.38	0.081642	0.74	0.010194	1.10	0.001437
0.03	0.976168	0.39	0.076789	0.75	0.009644	1.11	0.001362
0.04	0.933149	0.40	0.072247	0.76	0.009124	1.12	0.001291
0.05	0.876281	0.41	0.067995	0.77	0.008633	1.13	0.001224
0.06	0.813980	0.42	0.064010	0.78	0.008169	1.14	0.001160
0.07	0.751564	0.43	0.060276	0.79	0.007730	1.15	0.001100
0.08	0.691855	0.44	0.056774	0.80	0.007316	1.16	0.001043
0.09	0.636144	0.45	0.053488	0.81	0.006924	1.17	0.000989
0.10	0.584873	0.46	0.050405	0.82	0.006554	1.18	0.000937
0.11	0.538041	0.47	0.047510	0.83	0.006203	1.19	0.000889
0.12	0.495425	0.48	0.044790	0.84	0.005872	1.20	0.000843
0.13	0.456707	0.49	0.042235	0.85	0.005559	1.21	0.000799
0.14	0.421539	0.50	0.039833	0.86	0.005263	1.22	0.000758
0.15	0.389576	0.51	0.037575	0.87	0.004983	1.23	0.000718
0.16	0.360493	0.52	0.035451	0.88	0.004718	1.24	0.000681
0.17	0.333995	0.53	0.033453	0.89	0.004468	1.25	0.000646
0.18	0.309814	0.54	0.031573	0.90	0.004231	1.26	0.000613
0.19	0.287709	0.55	0.029803	0.91	0.004007	1.27	0.000581
0.20	0.267470	0.56	0.028136	0.92	0.003795	1.28	0.000551
0.21	0.248908	0.57	0.026566	0.93	0.003594	1.29	0.000522
0.22	0.231856	0.58	0.025088	0.94	0.003404	1.30	0.000496
0.23	0.216167	0.59	0.023695	0.95	0.003225	1.31	0.000470
0.24	0.201710	0.60	0.022382	0.96	0.003054	1.32	0.000446
0.25	0.188370	0.61	0.021145	0.97	0.002893	1.33	0.000423
0.26	0.176042	0.62	0.019978	0.98	0.002741	1.34	0.000401
0.27	0.164636	0.63	0.018878	0.99	0.002597	1.35	0.000380
0.28	0.154070	0.64	0.017841	1.00	0.002460	1.36	0.000361
0.29	0.144270	0.65	0.016862	1.01	0.002331	1.37	0.000342
0.30	0.135171	0.66	0.015939	1.02	0.002209	1.38	0.000325
0.31	0.126715	0.67	0.015068	1.03	0.002093	1.39	0.000308
0.32	0.118847	0.68	0.014246	1.04	0.001983	1.40	0.000292
0.33	0.111522	0.69	0.013470	1.05	0.001880	1.41	0.000277
0.34	0.104695	0.70	0.012738	1.06	0.001781	1.42	0.000263
0.35	0.098327	0.71	0.012046	1.07	0.001688	1.43	0.000249

Table 19.2 Smirnov–Cramèr–Von Mises Test (Continued)

$n\omega^2$	Signif	$n\omega^2$	Signifm	$n\omega^2$	Signifl	$n\omega^2$	Signif
1.44	0.000237	1.58	0.000114	1.72	0.000055	1.86	0.000026
1.45	0.000225	1.59	0.000108	1.73	0.000052	1.87	0.000025
1.46	0.000213	1.60	0.000102	1.74	0.000049	1.88	0.000024
1.47	0.000202	1.61	0.000097	1.75	0.000047	1.89	0.000023
1.48	0.000192	1.62	0.000092	1.76	0.000044	1.90	0.000021
1.49	0.000182	1.63	0.000087	1.77	0.000042	1.91	0.000020
1.50	0.000173	1.64	0.000083	1.78	0.000040	1.92	0.000019
1.51	0.000164	1.65	0.000079	1.79	0.000038	1.93	0.000018
1.52	0.000155	1.66	0.000075	1.80	0.000036	1.94	0.000017
1.53	0.000148	1.67	0.000071	1.81	0.000034	1.95	0.000017
1.54	0.000140	1.68	0.000067	1.82	0.000032	1.96	0.000016
1.55	0.000133	1.69	0.000064	1.83	0.000031	1.97	0.000015
1.56	0.000126	1.70	0.000061	1.84	0.000029	1.98	0.000014
1.57	0.000120	1.71	0.000058	1.85	0.000028	1.99	0.000013

$$\lim_{n\to\infty} P\left(D_{n_1,n_2} < \frac{\lambda}{\sqrt{n}}\right) = Q(\lambda),$$
$$n = n_1 n_2/(n_1 + n_2).$$
(19.6)

This is known as the Smirnov theorem. Note that this test is independent of the form of the experimental distribution.

Suppose we have k samples of n, $F^*_{n,1}(x)$, $F^*_{n,2}(x)$, ..., $F^*_{n,k}(x)$. Let $D(n, \; j)$ be the maximum absolute value of $F^*_{n,j}(x) - F(x)$, for all x and M_n be the maximum value of $D(n, \; j)$ for all j. Then, for a continuous distribution

$$P\left(M_n < \frac{\lambda}{\sqrt{n}}\right) = [Q\,(\lambda)]^k.$$
(19.7)

Sometimes we wish to test the relative difference, i.e., $(F^* - F)/F$ rather than the difference, $F^* - F$. We again suppose $F(x)$ is continuous except for discontinuity points. Let $R(a)$ be the maximum value of $(F^*(x) - F(x))/F(x)$ and $RA(a)$ be the maximum value of $|F^*(x) - F(x)|/F(x)$ for all x for which $F(x) \geq a$. It can then be shown (Renyi theorem) that for $\lambda > 0$

$$\lim_{n\to\infty} P\left(R(a) < \frac{\lambda}{\sqrt{n}}\right) = \sqrt{\frac{2}{\pi}} \int_0^{\lambda\sqrt{a/(1-a)}} \exp(-\lambda'^2/2)\, d\lambda' = T(\lambda, \; a),$$
(19.8)

Table 19.3 Kolmogorov–Smirnov λ-Distribution

λ	$Q(\lambda)$	λ	$Q(\lambda)$	λ	$Q(\lambda)$	λ	$Q(\lambda)$	λ	$Q(\lambda)$	λ	$Q(\lambda)$
0.32	0.0000	0.68	0.2558	1.04	0.7704	1.40	0.9603	1.76	0.9959		
0.33	0.0001	0.69	0.2722	1.05	0.7798	1.41	0.9625	1.77	0.9962		
0.34	0.0002	0.70	0.2888	1.06	0.7889	1.42	0.9646	1.78	0.9965		
0.35	0.0003	0.71	0.3055	1.07	0.7976	1.43	0.9665	1.79	0.9967		
0.36	0.0005	0.72	0.3223	1.08	0.8061	1.44	0.9684	1.80	0.9969		
0.37	0.0008	0.73	0.3391	1.09	0.8143	1.45	0.9702	1.81	0.9971		
0.38	0.0013	0.74	0.3560	1.10	0.8223	1.46	0.9718	1.82	0.9973		
0.39	0.0019	0.75	0.3728	1.11	0.8300	1.47	0.9734	1.83	0.9975		
0.40	0.0028	0.76	0.3896	1.12	0.8374	1.48	0.9750	1.84	0.9977		
0.41	0.0040	0.77	0.4064	1.13	0.8445	1.49	0.9764	1.85	0.9979		
0.42	0.0055	0.78	0.4230	1.14	0.8514	1.50	0.9778	1.86	0.9980		
0.43	0.0074	0.79	0.4395	1.15	0.8580	1.51	0.9791	1.87	0.9982		
0.44	0.0097	0.80	0.4559	1.16	0.8644	1.52	0.9803	1.88	0.9983		
0.45	0.0126	0.81	0.4720	1.17	0.8706	1.53	0.9815	1.89	0.9984		
0.46	0.0160	0.82	0.4880	1.18	0.8765	1.54	0.9826	1.90	0.9985		
0.47	0.0200	0.83	0.5038	1.19	0.8823	1.55	0.9836	1.91	0.9986		
0.48	0.0247	0.84	0.5194	1.20	0.8878	1.56	0.9846	1.92	0.9987		
0.49	0.0300	0.85	0.5347	1.21	0.8930	1.57	0.9855	1.93	0.9988		
0.50	0.0361	0.86	0.5497	1.22	0.8981	1.58	0.9864	1.94	0.9989		
0.51	0.0428	0.87	0.5645	1.23	0.9030	1.59	0.9873	1.95	0.9990		
0.52	0.0503	0.88	0.5791	1.24	0.9076	1.60	0.9880	1.96	0.9991		
0.53	0.0585	0.89	0.5933	1.25	0.9121	1.61	0.9888	1.97	0.9991		
0.54	0.0675	0.90	0.6073	1.26	0.9164	1.62	0.9895	1.98	0.9992		
0.55	0.0772	0.91	0.6209	1.27	0.9206	1.63	0.9902	1.99	0.9993		
0.56	0.0876	0.92	0.6343	1.28	0.9245	1.64	0.9908	2.00	0.9993		
0.57	0.0987	0.93	0.6473	1.29	0.9283	1.65	0.9914	2.01	0.9994		
0.58	0.1104	0.94	0.6601	1.30	0.9319	1.66	0.9919	2.02	0.9994		
0.59	0.1228	0.95	0.6725	1.31	0.9354	1.67	0.9924	2.03	0.9995		
0.60	0.1357	0.96	0.6846	1.32	0.9387	1.68	0.9929	2.04	0.9995		
0.61	0.1492	0.97	0.6964	1.33	0.9418	1.69	0.9934	2.05	0.9996		
0.62	0.1633	0.98	0.7079	1.34	0.9449	1.70	0.9938	2.06	0.9996		
0.63	0.1778	0.99	0.7191	1.35	0.9478	1.71	0.9942	2.07	0.9996		
0.64	0.1927	1.00	0.7300	1.36	0.9505	1.72	0.9946	2.08	0.9997		
0.65	0.2080	1.01	0.7406	1.37	0.9531	1.73	0.9950	2.09	0.9997		
0.66	0.2236	1.02	0.7508	1.38	0.9557	1.74	0.9953	2.10	0.9997		
0.67	0.2396	1.03	0.7608	1.39	0.9580	1.75	0.9956	2.11	0.9997		

Table 19.4 Kolmogorov–Smirnov λ-Distribution (Continued)

λ	$Q(\lambda)$	λ	$Q(\lambda)$	λ	$Q(\lambda)$	λ	$Q(\lambda)$	λ	$Q(\lambda)$
2.12	0.9998	2.16	0.9998	2.20	0.9999	2.24	0.9999	2.28	0.9999
2.13	0.9998	2.17	0.9998	2.21	0.9999	2.25	0.9999	2.29	0.9999
2.14	0.9998	2.18	0.9999	2.22	0.9999	2.26	0.9999	2.30	0.9999
2.15	0.9998	2.19	0.9999	2.23	0.9999	2.27	0.9999	2.31	1.0000

$$\lim_{n \to \infty} P\left(RA(a) < \frac{\lambda}{\sqrt{n}}\right)$$
$$= \frac{4}{\pi} \sum_{j=0}^{\infty} (-1)^j \exp\left(\frac{-([2j+1]^2\pi^2/8)(1-a)/2\lambda^2}{2j+1}\right) \tag{19.9}$$

If we wish to make a similar test to check if two independent samples are from the same distribution, we can use Wang's theorem. Suppose our two sample distribution functions $F_{n_1}^*(x)$, $F_{n_2}^*(x)$ have $n_1/n_2 \to d \leq 1$ as n_1, $n_2 \to \infty$. Let $R_{n_1,n_2}(a)$ be the maximum value of $(F_{n_2}^*(x) - F_{n_1}^*(x))/F_{n_2}^*(x)$ for all values of x for which $F_{n_2}^*(x) \geq a$. Then, for $\lambda > 0$

$$\lim_{n_2 \to \infty} P\left(R_{n_1,n_2}(a) < \frac{\lambda}{\sqrt{n}}\right) = T(\lambda, a), \tag{19.10}$$

$$n = n_1 n_2/(n_1 + n_2).$$

This test is also independent of the form of the experimental distribution.

19.3 Tests Based on the Binomial Distribution

Consider Fig. 19.1 again. If the hypothesis is a good one, then each bin has an independent probability of $\frac{1}{2}$ of falling above or below the hypothesized distribution. (We imagine the theoretical distribution is not arbitrarily normalized to force it to give the same number of events as the experimental one.) We can then turn to theorems on runs in coin tossing given in Feller (1950). We will only indicate the general idea here.

Let k be the number of runs of the same sign. Thus, $+ - - + + - - - +$ has $k = 5$, i.e., $+| - -| + +| - - -| + |$. If there is a long term trend in the data not guessed at in the hypothesis, then k will be small. If there is small scale jitter, k will be large. See Fig. 19.3.

Suppose the hypothesis is correct. Let n_1 be the number of $+$ bins, n_2 the number of $-$ bins, and $n = n_1 + n_2$ be the total number of bins in our histogram or points in our plot. Let $q(k)$ be the probability of k runs. We can show

Fig. 19.3 Sign changes between hypothesized and actual frequency functions. In case a, the data would tend to have too few sign changes and, in case b, too many sign changes compared with expectation for the hypothesized distribution

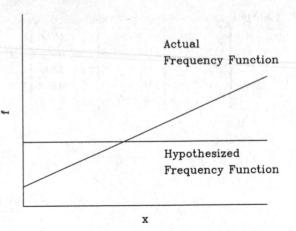

a. Too Few Sign Changes

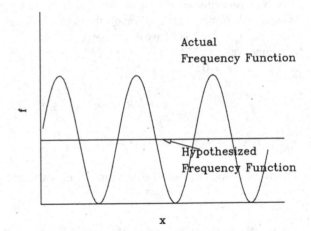

b. Too Many Sign Changes

$$q(k = 2v) = \frac{2 \binom{n_1 - 1}{v - 1} \binom{n_2 - 1}{v - 1}}{\binom{n}{n_1}} \qquad (19.11)$$

$$q(k = 2v + 1) = \frac{\binom{n_1 - 1}{v} \binom{n_2 - 1}{v - 1} + \binom{n_1 - 1}{v - 1} \binom{n_2 - 1}{v}}{\binom{n}{n_1}} \qquad (19.12)$$

Furthermore, the number of + signs should have a binomial distribution $P(n_1)$ with individual $p = \frac{1}{2}$

$$P(n_1 | p = \frac{1}{2}) = \binom{n}{n_1} \left(\frac{1}{2}\right)^n. \tag{19.13}$$

19.4 Tests Based on the Distributions of Deviations In Individual Bins of a Histogram

Consider Δ_i = number expected − number found in the ith bin. We ask, what is the distribution of $|\Sigma \Delta_i| \equiv \delta$, given $|\Delta_i|$ for all i? That is, we ask for the probability of the observed overall deviation given the values of the individual deviations without signs. There are 2^{n-1} possible values for δ each with probability $\left(\frac{1}{2}\right)^{n-1}$. A 1% test would be to reject the hypothesis if δ falls within the top 1% of the possible values (usually as determined numerically with a computer in practice). This is called a randomization test.

We can also examine the distribution of sizes of the deviations Δ_i. If we have a histogram with many events per bin, $\Delta_i / \sqrt{n_i}$ should approximately be distributed in a normal distribution (0,1). A plot of this quantity can be tested to see if it is normal in a standard manner with a least squares test or the integral distribution can be tested as discussed in the first section of this chapter.

Next, suppose we have two samples and wish to use this sort of test to see whether they are from the same distribution. Specifically, let us try to choose between the following two possibilities.

Hypothesis. The two samples are drawn from the same distribution.

Alternate. The median has shifted, but, otherwise, they are drawn from the same probability distribution.

Suppose our two samples are $x_1, x_2, \ldots, x_m$ and $y_1, y_2, \ldots, y_n$. Let V be the number of pairs (x_i, y_j) with $y_j > x_i$. For large m, n, V is distributed approximately in a normal distribution (Fraser 1960), with

$$E\{V\} = \frac{mn}{2}, \tag{19.14}$$

$$\text{variance}(V) = \frac{mn(m+n+1)}{12}. \tag{19.15}$$

Tables up to $m + n = 8$ are given by Mann and Whitney (1947). Tables in terms of a related statistic $T = V + [n(n+1)]/2$ are given up to $m + n = 20$ by Wilcoxon (1947). Similar tests are discussed by Fisz (1967).

Suppose in the above example we examine $D_y = \overline{y} - w$, where w is the combined sample mean. We can calculate D_y for each way of dividing the sample into m, n subsamples. There are $\binom{m+n}{n}$ ways of doing this. We can then test whether our observed D_y is near an extreme value or not. This is somewhat tedious, but for moderate values, it can be histogramed on a computer. As the sample size increases, it can be shown (Fraser 1960) that this is equivalent to the t test.

One test that has been suggested to see whether a peak in a histogram is real (Cowan 2018; Gross and Vitells 2010) is the following. Suppose you are hunting for a peak in a histogram and you have a test statistic q_0 defined so disagreement with data increases as q_0 increases. Suppose the location of the peak position is θ, defined only in the model in which there is a peak. An approximation for a global p-value can be shown to be $p_{global} = p_{local} + N_u$, where u is the value found for q_0 in your experiment and N_u (usually much smaller than one) is the mean number of "up-crossings" of q_0 above u in the range of the histogram. Davis (1987) finds that this number varies exponentially with the size of u, i.e. $N_u \approx N_{u_0} e^{-(u-u_0)/2}$. To the extent this is correct, one can extrapolate from a low value of u, which can reduce considerably the amount of Monte Carlo computing needed.

When should the methods in this chapter be used in preference to those described in previous chapters? The methods in this chapter tend to be robust, not sensitive to the form of the distribution. We need to look carefully at the data and at what we expect the distributions to be. If the errors are normal, then the previous methods should be satisfactory, although we may still wish to use the resolving power coming from looking at the integral probability distribution. A good rule here as well as for all applications is to not abandon common sense. Do not proceed blindly with a batch of rules. At each stage, ask if the results seem reasonable. Question. Question. Question.

19.5 Exercises

19.1 Verify that Eq. 19.2, does follow from Eq. 19.1, if $K(x) = F(x)$. Hint: Look at the equations for sums of integers given as a hint in Exercise 13.3.

19.2 When the 1987 Supernova (SN1987A) occurred, neutrino interactions from neutrinos made in the supernova were seen by the IMB experiment and by the Kamiokande experiment. Most of the events in the detectors were expected to produce recoil particles with an angular distribution of the form $d\sigma/d(\cos\theta) \propto (1 + \alpha \cos\theta)$ with respect to the direction of the supernova, with α in the range 0.05–0.15. Eight events were seen by the IMB group. The angles were 80, 44, 56, 65, 33, 52, 42, and 104 degrees. There were four Kamiokande events above the IMB threshold. They were at angles of 18, 32, 30, and 38 degrees. (The other KAM events were below the IMB energy threshold.) The experiments were sensitive to events in essentially the entire angular range $-1 < \cos\theta < 1$.

Examine the probability that these events come from the above distribution using the the Kolmogorov–Smirnov test and the Smirnov–Cramèr–Von Mises test. (The angular errors in the above are in the 10°–20° range and do not qualitatively affect the result.) As you see the probability is disturbingly low. Taking out the worst event (IMB 8) leaves the probability still small and including a small νe elastic scattering component does not help sufficiently either. The IMB group suggests there may be something new. Apparently, we must wait for the next nearby supernova in order to repeat the experiment!

References

Akerlof C (1991) Private communication; calculated the table of values for kolmogorov-smirnov tests

Anderson T, Darling D (1952) Ann Math Statist 23:193–212

Cowan G (2018) Statistics, in the review of particle physics. Phys Rev D 98(39):1–41

Davis RB (1987) Biometrica 74:33

Feller W (1950) Probability theory and its applications, vol 1. Wiley, New York

Fisz M (1967) Probability Theory and Mathematical Statistics. 3rd edn, Third Printing. Wiley, New York

Fraser D (1960) Statistics: an introduction, second printing. Wiley, New York

Gross E, Vitells O (2010) Trial factors for the look elsewhere effect in high energy physics. Eur Phys J C70:525. arxiv.org/pdf/1005.1891

Kiefer J (1959) K-sample analogues of the kolmogorov-smirnov and cramer-v. mises tests. Ann Math Statist 30:420

Mann H, Whitney D (1947) On a test of whether one of two random variables is stochastically larger than the other. Ann Math Statist 18:50

Wilcoxon F (1947) Probability tables for individual comparisons by ranking methods. Biometrics 3:119

Chapter 20
Characterization of Events

Abstract Many experiments collect quite a bit of information from each event. Often it is desired to classify these events into several different classes. For example, in a neutrino experiment, it may be useful to collect neutrino events into ν_e charged current events, ν_μ charged current events and other events. The simplest method is to make a series of cuts on the data. This has the advantage of clarity, one can see what is happening with each cut. However, since there are often correlations between the various pieces of information, it may not be efficient. Modern methods, such as neural nets and boosted decision trees (BDT) take the correlations into account and are more efficient. They use samples of known events to train the network and then run the trained network on the data events.

This efficiency comes at a price. It is very important that the sample used for training is a good match for the data events. An example of a successful test occurred looking at the recognition of handwritten digits 0–9. A training sample for a sophisticated convolutional neural network was made up of 50,000 handwritten digits, and then a second sample of 50,000 handwritten events were used as the data sample. The results were spectacular. Only 0.21% of the digits were unable to be correctly classified. This was considerably better than human classifiers could do.

Often the sample of "known" events for training is artificial events generated by a model of the events. It is quite possible and has happened, that the network trained on an artifact of the artificial events and the results were badly biased. This should be kept in mind as the discussion of these new methods proceeds.

20.1 Neural Networks

An artificial neural net (ANN) consists of several planes of nodes, with possibly differing number of nodes in each plane. The nodes are known as perceptrons. An excellent book discusses neural nets in some detail (Nielsen 2015). The first plane is the input plane, the last the output plane, and the other planes are "hidden" planes.

© Springer Nature Switzerland AG 2020

B. P. Roe, *Probability and Statistics in the Physical Sciences*,
Undergraduate Texts in Physics,
https://doi.org/10.1007/978-3-030-53694-7_20

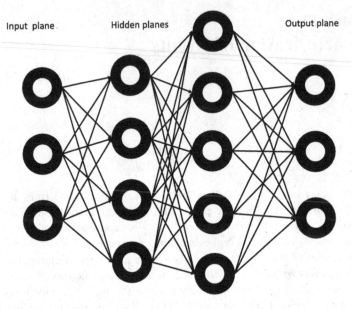

Fig. 20.1 Neural Net Diagram

The input plane has as inputs the various experimental data for an event. The output plane has as many nodes as the number of classes desired and outputs an appropriate number for each. Each node in each plane has connections to each node in the following plane (Fig. 20.1).

Suppose the output from node k is x_k. The input to a node j in the next plane contains a weight for each node in the input plane and a bias which depends on only node j, $z = w_{jk}x_k + b_j$. Usually, the input is chosen to be a function of these variables. A common choice is the sigma function, $\sigma(z) = 1/[1 + e^{-z}]$. Note that this sigmoid function goes from zero at $z = -\infty$ to one at $z = +\infty$. The net output a^l_j of node j in plane l from all nodes in the preceding plane, $l-1$, is

$$a^l_j = \sigma\left(\sum_k w^l_{jk}a^{l-1}_k + b^l_j\right) \text{ and } z^l_j = \sum_k w^l_{jk}a^{l-1}_k + b^l_j \qquad (20.1)$$

Suppose the desirable output from node i in the last plane L is y_i and x_i is the actual output. Think of $\mathbf{x}$, $\mathbf{y}$, $\mathbf{b}$, and $\mathbf{a}$ as column vectors. Define the cost function C, a measure of how far we are away from where we want to be as:

$$C = \frac{1}{2n}\sum_x ||\mathbf{y} - \mathbf{a}||^2 \qquad (20.2)$$

Here, $\mathbf{a}$ is the vector of all of the outputs (sigma functions) of the last plane, $||\mathbf{g}||$ is the length function of vector $\mathbf{g}$ and the sum over x is the sum over all n of the training events.

To help visualize this consider planes $l - 1$ and l. All nodes from $l - 1$ send their output to each node of l. Consider node k of $l - 1$ sending it's signal a_j^{l-1} to node j of l. First node j multiplies this signal by a weight and adds a bias. This produces a modified input $a_j^{l-1} w_{kj}^l + b_j^l$. The modified inputs from all the nodes of $l - 1$ are then added together to give an initial output z_j^l. The final output is then $a_j^l = \sigma(z_j^l)$. The initial inputs to plane 1 are the experimental quantities. It is often recommended that these be renormalized to be within the range 0 to $+1$. The final plane L has the same number of nodes j as the classifications desired (v_e charged current events, v_μ charged current events, other events). For training the net, each individual event is of a known kind and there is a desired output. For example, for a v_e inputted event, the desired output y_j might be 1 for that class and 0 for the other classes. The cost for one particular input event is then $(\sum_j (a_j^L - y_j)^2)/2$. The average of this result for the n input training events is then the overall cost C for this set of weights and biases.

20.1.1 Gradient Descent

It is necessary to try to minimize the cost. Suppose the cost is a function of m variables v, which are the various a's and b's and let $\mathbf{v}$ without subscript be the vector of these variables. The gradient of C is

$$\nabla C \equiv \left(\frac{\partial C}{\partial v_1}, \ldots, \frac{\partial C}{\partial v_m} \right)^T, \tag{20.3}$$

and $\Delta C \approx \nabla C \cdot \Delta \mathbf{v}$. To go down the slope to the minimum requires using minus the gradient. It is useful to do this slowly, so a factor η, the learning rate, is put in to go in small steps. η will have to be carefully chosen. η is often chosen as

$$\Delta \mathbf{v} = -\eta \nabla C. \tag{20.4}$$

A popular further choice is for $\eta = \epsilon/||\nabla C||$, where $||\nabla C||$ is the length function of ∇C. This means

$$||\Delta v|| = \epsilon. \tag{20.5}$$

ϵ is chosen to be reasonably small, sometimes in the region of 0.3, and sometimes a much lower value is needed, say 0.001. By repeatedly using this equation we can, hopefully, approach the minimum. This may take a long time if very small values of ϵ are needed. One way of shortening this is to use stochastic gradient descent. (SGD). Use a set of randomly chosen input samples and use the average ∇C of these samples.

An addition to gradient descent which is sometimes used is to add a term which changes the speed of going down hill, i.e., modifies the effective η This method is known as "momentum based gradient descent". Associate with each w_j variable a "velocity" v_j and modify Eq. 20.4, to be that for each step

$$v_j \to v'_j = \mu v_j - \eta \nabla C_j; \quad w_j \to w'_j = w_j + v'_j. \tag{20.6}$$

If each step keeps moving downhill, this adds to the size of the step each time. μ is usually chosen to be in the range zero to one. Values close to one moderate the increases in sizes of steps and help to prevent overshooting the minimum.

20.1.2 Back Propagation

Finding the appropriate derivatives of the cost with respect to all of the various weights and biases can be very complicated. However, there is a systematic way of handling this. It hinges on four main equations which will now be obtained. First the Hadamard product will be defined, as using it simplifies notation. Let $\mathbf{r}$ and $\mathbf{s}$ be two vectors of dimension n. The Hadamard product of the two vectors is the vector $\mathbf{H}$ whose elements are the products $r_i s_i$. This is written as $\mathbf{H} = \mathbf{r} \odot \mathbf{s}$. For the last plane L

$$\delta_j^L \equiv \frac{\partial C}{\partial z_j^L} = \sum_k \frac{\partial C}{\partial a_k^L} \frac{\partial a_k^L}{\partial z_j^L}$$

The sum is over all neurons in the last layer, but the output for layer k depends only on the input to layer k, so $j = k$ is the only non-zero term. Furthermore, $a_j^L = \sigma(z_j^L)$

$$\delta_j^L = \frac{\partial C}{\partial a_j^L} \sigma'(z_j^L) = \nabla_a C \odot \sigma'(z^L) \tag{20.7}$$

$$\delta_j^l \equiv \frac{\partial C}{\partial z_j^l} = \sum_k \frac{\partial C}{\partial z_k^{l+1}} \frac{\partial z_k^{l+1}}{\partial z_j^l} = \sum_k \frac{\partial z_k^{l+1}}{\partial z_j^l} \delta^{l+1}$$

where the definition of δ_k^{l+1} has been used. Next, note

$$z_k^{l+1} = \sum_j w_{kj}^{l+1} a_j^l + b_k^{l+1}.$$

$$\frac{\partial z_k^{l+1}}{\partial z_j^l} = w_{kj}^{l+1} \sigma'(z_j^l)$$

Substituting this back into the equation for δ_j^l gives

$$\delta_j^l = \sum_k w_{kj}^{l+1} \delta_k^{l+1} \sigma'(z_j^l) = ((w^{l+1})^T \delta^{l+1}) \odot \sigma'(z^l) \tag{20.8}$$

For the next two equations, recall $a_j^l = \sigma\left(\sum_k w_{jk}^l a_k^{l-1} + b_j^l\right)$ and $z_j^l = \sum_k w_{jk}^l a_k^{l-1} + b_j^l$. So

$$\frac{\partial z_j^l}{\partial b_j^l} = 1; \quad \frac{\partial C}{\partial b_j^l} = \frac{\partial C}{\partial z_j^l} \frac{\partial z_j^l}{\partial b_j^l} \quad \text{and} \quad \delta_j^l = \frac{\partial C}{\partial z_j^l}$$

$$\frac{\partial C}{\partial b_j^l} = \delta_j^l \tag{20.9}$$

$$\frac{\partial C}{\partial w_{jk}^l} = \frac{\partial C}{\partial z_j^l} \frac{\partial z_j^l}{\partial w_{jk}^l} \quad \text{and} \quad \frac{\partial C}{\partial w_{jk}^l} = a_j^{l-1}$$

$$\frac{\partial C}{\partial w_{jk}^l} = a_k^{l-1} \delta_j^l \tag{20.10}$$

Given these four Eqs. (20.7)–(20.10), we can then perform a gradient descent. First input the initial x values to the first plane and have a trial set of w's and b's. It is then easy to go forward to find C after the last plane. Then the second of the four equations enables us to go back a step and in succeeding steps the effect of all of the weights and biases on the cost can be calculated. Then the gradient descent method of the previous section can be followed to try to minimize the cost.

What should be the original choice of weights and biases? In the absence of knowledge from the experiment one choice might be to choose all weights and biases as normally distributed random numbers with mean zero and standard deviation one. This choice has a problem. If a node in an intermediate plane is fed by n nodes from the preceding plane, then, for that node $z = \sum_j w_j x_j + b$, initially has a standard deviation of $\sqrt{n+1}$, which is very large and σ is near saturation which will slow down the fitting. A better choice is to choose the standard deviation of the weight of each node to be $\sqrt{1/n}$.

20.1.3 Some Neural Net Variations

The cost function used above, $C = \frac{1}{2n} \sum_x ||y - a||^2$ can converge very slowly using gradient descent if the initial guesses for the weights and biases are far from the needed values. The basic problem is that the derivatives C necessarily involve derivatives of $\sigma(z) = 1/[1 + e^{-z}]$. If z is close to one or to zero, changes in its value produce only small changes in $\sigma(z)$. Several alternative cost functions and activation functions will be examined here,

A new cost function, called the cross-entropy cost function is

$$C = -\frac{1}{n}\sum_x \sum_j [y_j \ln a_j^L - (1 - y_j) \ln(1 - a_j^L)] \geq 0. \qquad (20.11)$$

Note that if $y = 0$ and $a \approx 0$ or if $y = 1$ and $a \approx 1$ then $C \to 0$. This avoids the problem of σ'. It can be shown that $\sigma'(z) = \sigma(z)(1 - \sigma(z))$ from which it is easy to show

$$\frac{\partial C}{\partial w_{jk}^L} = \frac{1}{n}\sum_x a_k^{L-1}(a_j^L - y_j) \qquad (20.12)$$

and the derivative σ' does not appear. Similarly, it follows that

$$\frac{\partial C}{\partial b_j}^L = \frac{1}{n}\sum_x (a_j^L - y_j) \qquad (20.13)$$

Another way to avoid slowdown in learning is to just use z_j^l as output, not $\sigma(z_j^l)$.

A special activation function for the output layer, known as the softmax function is

$$a_j^L = \frac{e^{z_j^L}}{\sum_k e^{z_k^L}}. \qquad (20.14)$$

Since $\sum_j a_j^L = 1$, the output can be thought of as a probability. The various channels are no longer independent. A change in one j channel affects all of the other channels. Equations 20.12 and 20.13 can be shown to hold for the softmax function which means it, also, would not suffer from slow convergence.

Using the training sample and gradient descent for, say m, cycles can sometimes lead to overfitting. If m gets to be too large, the cost may keep going down, but the actual accuracy on a test sample may plateau or even get worse. A standard way to proceed is to use a third sample, a verification sample. Look at the saturation in accuracy on the test sample. When it plateaus, then try the verification sample as an independent measure of accuracy. This is an important, but difficult problem. For example, it may happen that the accuracy plateaus for a while and then again starts to increase.

Another way to help with overfitting is to use a regularization technique as described in Chap. 15. This technique helps suppress noise by reducing high frequency noise. L2 regularizaton, sometimes called weight decay, adds a new term to the cost function C_0

$$C = C_0 + \frac{\lambda}{2n}\sum_w w^2; \quad \frac{\partial C}{\partial w_{jk}^l} = \frac{\partial C_0}{\partial w_{jk}^l} + \frac{\lambda}{n}w_{jk}^l \qquad (20.15)$$

where the sum is over all of the weights in the net. Note the biases are not included in the regularization term. λ is a constant set by hand. To see the effect for gradient descent, just take the derivatives of Eq. 20.15. Parameters such as λ, η and μ are known as hyper parameters. L1 regularization is similar to L2 regularization, but uses the magnitudes of $|w|$ not the w^2

$$C = C_0 + \frac{\lambda}{n} \sum_w |w|; \quad \frac{\partial C}{\partial w_{jk}^l} = \frac{\partial C_0}{\partial w_{jk}^l} + \frac{\lambda}{n} sgn(w_{jk}^l), \qquad (20.16)$$

where $sgn(w)$ is the sign of w. If $w = 0$, take $sgn(w) = 0$ Although these are somewhat ad hoc modifications, it is observed that regularization tends to reduce overfitting.

The choice of the initial value for hyper parameters need special care. Some heuristic methods for choosing them are given in Nielsen (2015), Bengio (2012). Some semi-automatic methods for choosing them are the use of Bayesian methods (Snoek et al. 2012) and a grid search (Bergstra and Bengio 2012).

Another method sometimes used to reduce overfitting is called dropout. Dropout consists of first randomly deactivating half of the nodes in the intermediate planes, leaving the input and output planes intact. Training is done several times, each with different random choices of nodes to dropout, each time starting with the weights and biases found in the previous dropout trial. This corresponds to testing different neural networks, reducing the interdependence of the various parameters. After this has been done several times, the full network is run. Since there are twice as many hidden nodes as in the dropout samples, the weight of each hidden node should be halved. This method leads to a set of parameters more robust to the loss of any particular node.

A remarkable theorem is that it can be shown that given an input plane with i input neurons x_i, an output plane with j output neurons a neural net can be built to fit as accurately as you wish a function f_j for any reasonably arbitrary function $f_k(x_1, x_2, \ldots, x_i)$ for $k = 1, 2, \ldots, j$ (Nielsen 2015).

It may be useful for neural nets and the boosted decision tree method to be discussed later, to find combinations of correlated, possibly non-normal, variables to obtain variables which have the highest discrimination for separating signal and background. The Fisher discriminant method suggests using linear combinations of variables which maximize $(\overline{y}_{sig} - \overline{y}_{bkg})^2/(\sigma_{sig}^2 + \sigma_{bkg}^2)$. Some expansions of this method are discussed in Cowan (1998), Roe (2003).

20.2 Deep Learning

Deep learning networks are just neural networks with several hidden planes. However, the presence of several hidden planes makes them difficult to tune. The basic problem is that different parts of the network train at very different rates. Schematically, if gradient descent is used, the change in C from a change in b at plane 1 through L is

$$\frac{\partial C}{\partial b_1} = \sigma'(z_1) \times w^2 \times \sigma'(z_2) \times \cdots \times w^L \times \sigma'(z_L) \times \frac{\partial C}{\partial a_L}$$

The maximum value of $\sigma'(z)$ is 0.25. If the weights and biases are reasonably small, then the change in plane 1 is greatly reduced and it learns at a very slow rate. This is known as the vanishing gradient. With some large weights and appropriate biases, the change in plane 1 can be much larger than that in the last plane. This is known as gradient explosion. The main point is the instability, that parts of the net may learn at much different rates than other parts. A solution is to reorganize the network connections, to use "convolutional neural networks" (LeCun et al. 1998). So far, the input has been a one-dimensional line of nodes. Suppose it is desired to enter the basic data from each event as a series of pixels, an image. This was originally designed for text or picture images, but is now being used in particle physics (Aurisano et al. 2016). For a two dimensional image, the input is an $n \times n$ input plane. It would be impractical to connect each pixel to each pixel in the next plane. Furthermore, usually, most of the information in obtaining an image is in nearby pixels. For a convolutional hidden layer, each node in the first hidden layer will be connected to only a small part of the input, say an $s \times s$ square, which is called the local receptive field. Start in the upper left corner and connect to the first node in the first hidden layer. Then slide over by one pixel and connect the second. Do this both horizontally and vertically, ending with $s \times s$ of the input cells connected to one node of the hidden layer. If the first layer has $t \times t$ nodes, then the hidden layer has $(t - s + 1) \times (t - s + 1)$ nodes. It is not necessary to move by only one pixel in sliding over. Stride lengths of more than one are often used. The weight and bias for each of the hidden nodes are chosen to be the same. Each node of the hidden node is then looking for the same feature, but at a different part of the image. The hidden layer is then called a feature map. The weights and biases are called shared weights and biases.

$$\sigma \left(b + \sum_{l=0}^{s-1} \sum_{m=0}^{s-1} w_{l,m} a_{j+l,k+m} \right), \tag{20.17}$$

where j and k refer to the $(t - s + 1) \times (t - s + 1)$ hidden nodes. σ is the neural activation function. The sigmoid, the softmax and the rectified linear functions are all often used here. The rectified linear function is just $max(0, z)$. It is often the case that it is desirable to look at several features f in the first hidden layer. That hidden layer then has dimension $s - 1 \times s - 1 \times f$. Because of the shared parameters, the number of parameters is greatly reduced over a normal neural net with connections from each node in each plane to each node in the following plane, which helps reduce the problem of finding minima. Another kind of hidden layer is the pooling layer, usually used right after the convolutional layer. In this plane, each node summarizes a small number of the nodes of the previous convolutional plane. For example, it might summarize a $p \times p$ region. There are various ways to summarize this information. One method is max-pooling, in which, the summary is just to take the maximum activation of the nodes to be summarized. Another method is L2 pooling which takes the square root of the sum of the squares of the activations. The summaries look for a particular feature, but give away some of the information about the location of the feature. If there are f features, each one is pooled separately. There may be several convolution and pooling layers. After the final convolutional and pooling layers, the final output layer is completely connected to the preceding layer. To use a gradient descent method, it is necessary to do back propagation.

Previously, every node in a layer was connected to every node in the next layer. That is not true here and the back propagation formulas must be modified accordingly. The regularization techniques discussed previously can be used here to assist the gradient descent. Another technique which can help is to augment the training set by using each event several times but each having the image translated or rotated somewhat.

20.2.1 Generative Adversarial Networks

Suppose that a convolutional neural network is trained to recognize that an object is a face. Given a training sample, the result is some average over the various faces on which it trained. If a specific face is put into the trained network the image that comes out can be blurred. "Generative adversarial networks" (GAN) (Goodfellow 2017; Goodfellow et al. 2016) can help reduce the problem. This method involves two functions. G is called the generator and D is the discriminator. G and D have to be differentiable functions. This is played out as a game between the two players. The generator creates samples that are intended to come from the data sample The discriminator looks at the samples to see whether they are real or fake. The generator is trying to make samples which look like data samples, and the discriminator divides samples into "real" or "fake". I. Goodfellow, who invented this method notes that the generator is trying to build samples of money that counterfeit the data, and the discriminator is like the police, trying to allow legitimate money and reject counterfeit money. G and D are usually convolutional neural networks. $G(\mathbf{z}, \theta^{(G)})$ has the vector $\mathbf{z}$ as input and parameters $\theta^{(G)}$. The parameters are the weights and biases of G. Similarly, $D(\mathbf{x}, \theta^{(D)})$ has the vector $\mathbf{x}$ as input and parameters $\theta^{(D)}$. Both networks have cost functions, $C^{(D)}(\theta^{(D)}, \theta^{(G)})$ and $C^{(G)}(\theta^{(D)}, \theta^{(G)})$. Although the cost of each depends on both sets of parameters, D and G each only have control over their own parameters. Optimization is obtained when a point in parameter space is found where all neighboring points have greater or equal costs. This means that C^D is a local minimum with respect to $\theta^{(D)}$ and C^G is a local minimum with respect to $\theta^{(G)}$. The inputs to G can be modified at any stage as desired, by adding noise, or adding a random normal variable to the last layer of the net. We do want the dimension of $\mathbf{z}$ to be at least as large as the dimension of $\mathbf{x}$.

The original input data ($\mathbf{z}$) to G is just noise, for example normally distributed random numbers with mean zero and variance one. For training simultaneous SGD is used, with mini samples $\mathbf{z}$ taken from the output of the previous run of G, and from the real data. The cost function for the discriminator is essentially cross-entropy

$$C^{(D)}(\theta^{(D)}, \theta^{(G)}) = -\frac{1}{2}\mathbb{E}_{\mathbf{x} \sim p_{\text{data}}}\{\log D(\mathbf{x})\} - \frac{1}{2}\mathbb{E}_{\mathbf{z}}\{\log(1 - D(G(\mathbf{z}))\}, \quad (20.18)$$

where $\mathbb{E}_{\mathbf{x}}$ means "expected value over $\mathbf{x}$". The cost for G is often chosen heuristically as

$$C^{(G)} = -\frac{1}{2}\mathbb{E}_{\mathbf{z}}\{\log D(G(\mathbf{z})\}. \quad (20.19)$$

The generator maximizes the log-probability of the discriminator being mistaken. In the fitting for G, back propagation goes from the output of D down to the first stage of G. When G updates are done, D parameters are held fixed. At the end, it is G that is used to give the representation of the image. Some suggestions for improving performance have been to set the target for D to 0.9, not 1.0. Another suggestion is to add noise to the data images. GAN's are a rapidly developing field. They are already being used in a number of physics applications. Various improvements are described in Goodfellow et al. (2016), Radford et al. (2015) and in a 2017 review (Creswell et al. 2017). Some new forms of neural nets are discussed in Radovic et al. (2018) and a recent review is given in Carleo et al. (2020).

20.3 Boosted Decision Trees

This is a different approach to classification. With convolutional neural networks, it was possible to start with almost the raw data for each event. Boosted decision trees (BDT) (Roe et al. 2005; Yang et al. 2005, 2007) were studied before the advent of convolutional neural networks. BDT's start the analysis after some processing of the event. There are both advantages and disadvantages to this approach. The neural network can give excellent results, but it is sometimes hard to know how it has done it. This can lead to problems if the net has focussed on an artifact of a Monte Carlo input. For the BDT, one starts with more physical parameters, and can keep track of which parameters were the most important for distinguishing signal from noise. However, just as with ANN, it is very important that the parameters of the training samples are a good match to the data. The speed of training needed for BDT's is usually faster than that for ANN's.

Suppose one is trying to divide events into signal and background and suppose Monte Carlo samples of each are available. Divide each Monte Carlo sample into two parts. The first part, the training sample, will be used to train the decision tree, and the second part, the test sample, to test the final classifier after training. Since the test sample is used to see how the training is proceeding, a third sample, the verification sample, is useful, just as it was for neural nets. For each event, suppose there are a number of particle identification (PID) variables useful for distinguishing between signal and background. Firstly, for each PID variable, order the events by the value of the variable. Then pick variable one and for each event value see what happens if the training sample is split into two parts, left and right, depending on the value of that variable. Pick the splitting value which gives the best separation into one side having mostly signal and the other mostly background. Then repeat this for each variable in turn. Select the variable and splitting value which gives the best separation. Initially, there was a sample of events at a parent "node". Now there are two samples called child "branches". For each branch, repeat the process, i.e., again try each value of each variable for the events within that branch to find the best variable and splitting point for that branch. One keeps splitting until a given number of final branches, called "leaves", are obtained, or until each leaf is pure signal or pure background, or has too few events to continue. This description is a little oversimplified. In fact, at each stage, one picks as the next branch to split, the branch which will give the best increase in the quality of the separation. What criterion is used to define the quality of separation between signal and background in the split? Imagine the events are weighted with each event having weight w_i. Define the purity of the sample in a branch by

$$P = \frac{\sum_s w_s}{\sum_s w_s + \sum_b w_b} \tag{20.20}$$

Here w_s is the weight of a signal event and w_b the weight of a background event. Note that $P(1 - P)$ is zero if the sample is a pure signal or pure background. For a given branch, let

$$Gini = \sum_i^n w_i P(1 - P), \tag{20.21}$$

where n is the number of events on that branch. The criterion chosen is to minimize

$$Gini_{left\ child} + Gini_{right\ child}$$

To determine the increase in quality when a node is split into two branches, one maximizes

$$C = \text{Criterion} = Gini_{parent} - Gini_{child\ left} - Gini_{child\ right} \tag{20.22}$$

At the end, if a leaf has a purity greater than 1/2 (or whatever is set), then it is called a signal leaf and if the purity is less than 1/2, it is a background leaf. Events are classified signal if they land on a signal leaf and background if they land on a background leaf. The resulting tree is a "decision tree" (Breiman et al. 1984).

There are three major measures of node impurity used in practice: misclassification error, the Gini index, and the cross-entropy. If we define p as the proportion of the signal in a node, then the three measures are: $1 - max(p, 1 - p)$ for the misclassification error, $2p(1 - p)$ for the Gini index and $-p\log(p) - (1 - p)\log(1 - p)$ for the cross-entropy. The three measures are similar, but the Gini index and the cross-entropy are differentiable, and hence more amenable to numerical optimization. In addition, the Gini index and the cross-entropy are more sensitive to change in the node probabilities than the misclassification error.

A considerable improvement to the method occurred when the concept of boosting was introduced (Schapire 2002; Freund and Schapire 1999; Friedman 2003). Start with unweighted events and build a tree as above. If a training event is misclassified, i.e, a signal event lands on a background leaf or a background event lands on a signal leaf, then the weight of that event is increased (boosted).

A second tree is built using the new weights, no longer equal. Again misclassified events have their weights boosted and the procedure is repeated. Typically, one may build 1000 or 2000 trees this way. The number of branches in the tree can be reasonably large. I have used 50 branches with no problem.

A score is now assigned to an event as follows. The event is followed through each tree in turn. If it lands on a signal leaf it is given a score of 1 and if it lands on a background leaf it is given a score of -1. The renormalized sum of all the scores, possibly weighted, is the final score of the event. High scores mean that the event is most likely signal and low scores that it is most likely background. By choosing a particular value of the score on which to cut, one can select a desired fraction of the signal or a desired ratio of signal to background. It works well with many PID variables. Some trees have been built with several hundred variables. If one makes a monotonic transformation of a variable, so that if $x_1 > x_2$ then $f(x_1) > f(x_2)$, the boosting method gives exactly the same results. It depends only on the ordering according to the variable, not on the value of the variable.

Next, turn to methods of boosting. If there are N total events in the sample, the weight of each event is initially taken as $1/N$. Suppose that there are N_{tree} trees and m is the index of an individual tree. Let

$x_i =$ the set of PID variables for the ith event.

$y_i = 1$ if the ith event is a signal event and $y_i = -1$ if the event is a background event.

$T_m(x_i) = 1$ if the set of variables for the ith event lands that event on a signal leaf and $T_m(x_i) = -1$ if the set of variables for that event lands it on a background leaf.

$I(y_i \neq T_m(x_i)) = 1$ if $y_i \neq T_m(x_i)$ and 0 if $y_i = T_m(x_i)$.

There are at least two commonly used methods for boosting the weights of the misclassified events in the training sample. The first boosting method is called AdaBoost (Freund and Schapire 1999). Define for the mth tree:

$$err_m = \frac{\sum_{i=1}^{N} w_i I(y_i \neq T_m(x_i))}{\sum_{i=1}^{N} w_i} \qquad (20.23)$$

$$\alpha_m = \beta \times \ln((1 - err_m)/err_m) \qquad (20.24)$$

$\beta = 1$ is the value used in the standard AdaBoost method. I have found $\beta = 0.5$ to give better results for my uses. Change the weight of each event $i = 1, \cdots, N$

$$w_i \rightarrow w_i \times e^{\alpha_m I(y_i \neq T_m(x_i))}. \tag{20.25}$$

Each classifier T_m is required to be better than random guessing with respect to the weighted distribution upon which the classifier is trained. Thus, err_m is required to be less than 0.5, since, otherwise, the weights would be updated in the wrong direction. Next, renormalize the weights, $w \rightarrow w/N_w$. The score for a given event is

$$T(x) = \sum_{m=1}^{N_{trees}} \alpha_m T_m(x), \tag{20.26}$$

which is just the weighted sum of the scores over the individual trees. The second boosting method is called ϵBoost (Friedman 2003) or sometimes "shrinkage". After the mth tree, change the weight of each event $i = 1, \ldots, N$

$$w_i \rightarrow w_i e^{2\epsilon I(y_i \neq T_m(x_i))} \tag{20.27}$$

where ϵ is a constant of the order of 0.01. Renormalize the weights, $w_i \rightarrow w_i/(\sum_{i=1}^{N} w_i)$. The score for a given event is

$$T(x) = \sum_{m=1}^{N_{trees}} \epsilon T_m(x), \tag{20.28}$$

which is the renormalized, but unweighted, sum of the scores over individual trees. The AdaBoost and ϵBoost algorithms try to minimize the expectation value: $\mathbb{E}(e^{yF(x)})$, where $y = 1$ for signal and $y = -1$ for background. $F(x) = \sum_{m=1}^{N_{trees}} f_m(x)$, where the classifier $f_m(x) = 1$ if an event lands on a signal leaf and $f_m(x) = -1$ if the event falls on a background leaf. This minimization is closely related to minimizing the binomial log-likelihood (Friedman 2001). It can be shown that $\mathbb{E}\{e^{-yF(x)}\}$ is minimized at

$$F(x) = \frac{1}{2} \ln \frac{P(y = 1|x)}{P(y = -1|x)} = \frac{1}{2} \ln \frac{p(x)}{1 - p(x)}. \tag{20.29}$$

Let $y^* = (y + 1)/2$. It is then easy to show that

$$e^{-yF(x)} = \frac{|y^* - p(x)|}{\sqrt{p(x)(1 - p(x))}}. \tag{20.30}$$

The right-hand side is known as the χ statistic. χ^2 is a quadratic approximation to the log-likelihood, so χ can be considered a gentler alternative. It turns out that fitting using χ is monotone and smooth; the criteria will continually drive the estimates towards purer solutions. As noted in the neural net discussion, an ANN often tries to minimize the squared-error $\mathbb{E}\{(y - F(x))^2\}$, where $y = 1$ for signal events, $y = 0$ for background events, and $F(x)$ is the network prediction for training events. In Yang et al. (2005) several more comparisons were made between Adaboost, ϵBoost and several others. ϵBoost converges more slowly than does AdaBoost; however, with about 1000 tree iterations, their performances are very comparable and better than the other methods that were tried. Removing correlations of the input PID variables was found to improve convergence speed. In Yang et al. (2007) it was also found that the degradation of the classifications obtained by shifting or smearing variables of testing results is smaller for BDT than for regular ANN's.

It has been suggested that it might be useful to use only a randomly chosen sample of the events for each tree and, perhaps, a randomly chosen fraction of the variables for each node within each tree. It is useful, in practice to print out the purity perhaps once every 25 or 50 trees. It is also useful to keep track of how many times each PID variable is used as the cutting variable and some measure of the overall effectiveness of that variable. That will indicate which variables were the most important of the PID variables.

There have been a number of suggestions for improvements in BDT methods for special purposes. For example, a recent BDT method labeled QBDT is claimed to be able to take systematic uncertainties into account (Xia 2019).

20.4   Exercises

20.1 Prove the relation, $\sigma'(z) = \sigma(z)(1 - \sigma(z))$ that was used in the discussion of the cross-entropy cost function.

20.2 Use CERN's boosted decision tree program, and the data given at the website https://archive.ics.uci.edu/ml/datasets/MiniBooNE+particle+identification to separate out signal (ν_e events) from background. Compare with the results obtained in (Roe et al. 2005). A brief description of the variables is given in https://www-boone.fnal.gov/tech_notes/tn023.pdf.

20.3 Use the 12 or so most used BDT variables determined in problem 2 and use CERN's neural net program to try the same selection. Compare with the neural net results in Roe et al. (2005). Since neural net understanding has improved since 2005, you may find you can improve over the results presented in the reference.

References

Aurisano A, Radovic A, Rocco D, et al (2016) A convolutional neural network neutrino event classifier. arxiv.org/pdf/1604.01444
Bengio Y (2012) Practical recommendations for gradient-based training of deep architectures. arxiv.org/pdf/1206.5533v2

Bergstra J, Bengio Y (2012) Random search for hyper-parameter optimization. J Mach Learn Res

Breiman L, Friedman J, Olshen C, Stone RA (1984) Classification and regression trees. Wadsworth International Group, Belmont, California

Creswell A, White T, Dumoulin V, Arulkumaran K, et al (2017) Generative adversarial networks: An overview. submitted to IEEE-SPM arxiv.org/pdf/1710.07

Cowan G (1998) Statistical data analysis. Oxford Clarendon Press, Oxford

Carleo G et al (2020) Rev Mod Phys 91(045):002

Freund Y, Schapire RE (1999) A short introduction to boosting. J Jpn Soc Artif Intell (Appearing in Japanese, translation by Abe, N) 14(5):771–780

Friedman J (2003) Recent advances in predictive (machine) learning. In: Proceedings of Phystat 2003, Stanford U

Friedman J (2001) Greedy function approximation: a gradient boosting machine. Ann Stat 29(5):1189–1232

Goodfellow I (2017) Nips tutorial: generative adversarial networks. arxiv.org/pdf/1701.00160v4

Goodfellow I, Bengio Y, Courville A (2016) Deep learning. http://www.deeplearningbook.org

LeCun Y, Bottou L, Bengio Y, Hafner P (1998) Gradient-based learning applied to document recognition. https://ieeexplore.ieee.org/document/726791

Nielsen M (2015) Neural nets and deep learning

Radovic A, Williams M, Rousseau D et al (2018) Machine learning at the intensity frontiers of particle physics. Nature 560:41–48

Roe B (2003) Event selection using an extended fisher discriminant method. arxiv.org/pdf/0310145

Radford A, Metz L, Chintala S (2015) Unsupervised representation learning with deep convolutional generative adversarial networks. arxiv.org/pdf/1511.06434

Roe BP, Yang HJ, Zhu J, et al (2005) Boosted decision trees as an alternative to artificial neural networks for particle identification. Nucl Instrum Meth A543:577–584. arxiv.org/pdf/0408124

Schapire RE (2002) The boosting approach to machine learning: an overview. In: MSRI workshop on nonlinear estimation and classification

Snoek J, Larochelle H, Adams R (2012) Practical bayesian optimization of machine learning algorithms. arxiv.org/pdf/1206.2944

Xia LG (2019) Qbdt, a new boosting decision tree method with systematical uncertainties into training for high energy physics. arxiv.org/pdf/1810.08307v4

Yang HJ, Roe BP, Zhu J (2005) Studies of boosted decision trees for miniboone particle identification. Nucl Instrum Meth A555:370–385. arxiv.org/pdf/0508045

Yang HJ, Roe BP, Zhu J (2007) Studies of stability and robustness for artificial neural networks and boosted decision trees. Nucl Instrum Meth A574:342–349. arxiv.org/pdf/0610276

Appendix A
Brief Notes on Getting Started with CERN Root

The first thing that is needed is to download root onto your computer:

Go to https://root.cern.ch/downloading-root where instructions for downloading root and setting it up on your computer are given. Downloading root using binaries: After unpacking/installing the binary, and before using ROOT you should use a special script distributed with ROOT:

source <pathname>/root/bin/thisroot.sh (there are versions for [t]csh and fish, too.) where <pathname> is the location where you unpacked the ROOT distribution. Typically, add these lines to your .profile or .login files. The same location also leads to a Reference Guide for using root.

Root is typically run by adding a C fragment, say Plot.c. It can be called either by "root -q -b Plot.c" or by

root

.x Plot.c

In either case ".q" will end root

Some simple fragments for plotting or fitting data to a curve or indicated below. The plot fragment assumes that there are tables stored in the file test.dat. This file is shown below (Fig. A.1)

```
304.9  803000.0
354.9  581083.0
404.9  429666.0
454.9  320016.0
504.9  242783.0
554.9  188487.0
604.9  150737.0
654.9  124103.0
704.9  105397.0
754.9   92748.0
```

The Plot.c file to use for this is

© Springer Nature Switzerland AG 2020
B. P. Roe, *Probability and Statistics in the Physical Sciences*,
Undergraduate Texts in Physics,
https://doi.org/10.1007/978-3-030-53694-7

```
//for discussion of graphs in root see
// https://root.cern.ch/root/html534/guides/users-guide/Graphs.html
gROOT->SetStyle("Plain");
gStyle->SetPalette(1,0);
gStyle->SetLineWidth(1);
gStyle->SetHistLineWidth(2);
gStyle->SetOptFit(kTRUE);
gStyle->SetOptFit(111);
gStyle->SetOptStat(0);
gStyle->SetEndErrorSize(3);
double time, counts;
double A =6110170.;
double lam= .00702929.;
double B =60969.5.;
double x[500000];
double y[500000];
double z[500000];
double t[500000];
double errx[50000];
double errt[50000];
long int i=0;
string line;
ifstream infile;
infile.open("test.dat");
if (infile.is_open())
while ( getline (infile,line)  )
istringstream ss(line);
ss » time » counts;
t[i]=time;
x[i]=counts;
z[i] =A*exp(-lam*time)+B;
errx[i] = sqrt(z[i]);
// the errors here are too small to see.  Try mult errx by 50.
errt[i] = 0.;
i++;
infile.close();
gr = new TGraph(i,t,x);
grr = new TGraph(i,t,z);
new TCanvas;
//gPad->GetCanvas()->SetLogx(); This sets a logarithmic x-axis
gPad->GetCanvas();
gr->GetYaxis()->SetTitleOffset(1.28);
gr->GetXaxis()->SetTitle("Time (ns)");
gr->GetYaxis()->SetTitle("Counts");
gr->SetTitle("Counts vs Time (ns) for radiative decay");
```

```
gr->GetXaxis()->SetRangeUser(0.,800.);
gr->GetYaxis()->SetRangeUser(90000.,1100000.);
gr->SetLineColor(1);
gr->SetLineStyle(1);
//gr->Draw("ALP");
//A means axes around graph,L means polyline,
// P current marker drawn at every point, * means marker is a star
gr->Draw("AP*");
grr = new TGraphErrors(10,t,x,errt,errx);
//This means t and x errors will be drawn
grr->SetLineStyle(1);
grr->Draw("SAME"); // Use the same graph; don't start new graph
// TLegend *leg = new TLegend(0.65, 0.75, 0.75, 0.9);
//TLegend *leg = new TLegend(0.2, 0.4, 0.4, 0.6);
//leg->AddEntry(gr,"mu-e", "L");
//leg->AddEntry(grr,"mu-mu", "L");
//leg->AddEntry(grrr,"mu-tau", "L");
//leg->SetFillColor(0);
//leg->SetBorderSize(0);
//leg->SetTextSize(0.03);
//leg->Draw("same");
gPad->GetCanvas()->Print("test.pdf");
```

Fig. A.1 Figure from Plot.c

In MINUIT, because of the way the program tries to avoid recalculating the error matrix G, the errors printed as the minimization proceeds may not correspond to a newly calculated G^{-1}. Furthermore, the G's correspond only to the best estimates at that time and may not be very good at all. After a minimum is found, the error matrix should be reevaluated. (In MINUIT, this is done using the routine HESSE, which calculates the errors by recalculating and inverting the second derivative matrix, or with MINOS, which calculates the errors by tracing the contours of the function.)

The next Plot.c shows how to use MINUIT

```
//Define the data we want to fit - 11 (x,y) points and the y-errors
double time[10];
double x[10];
double timetemp;
double xtemp;
//double z[10];
string line;
double A =6110170.;
double lam= .00702929.;
double B =60969.5.;
//Define the function to minimize
//The form of the function is controlled by Minuit
void FCN(Int_t &npar, Double_t *gradIn, Double_t &funcVal, Double_t *par,
Int_t flag) {
//put the fitting function here.  It must come before void prob15_1minuit()
//Here the total number of events is not fixed if chi2.  Therefore use
maximum likelihood
//  Double_t chi2 = 0;
Double_t xlike = 1.;
for (int i = 0; i < 10; ++i) {
Double_t delta = x[i] - par[0]*exp(-par[1]*time[i])-par[2];
xlike = xlike +(delta*delta)/(2.*x[i]) +.5*log(x[i]);
}
funcVal = xlike;
}
//Load the data and put it into a histogram–TH1D defines histogram
//if you wish to histogram the result–10 bins,
//time from 304.9. to 754.9. in equal intervals
//Since equal intervals, we can ignore the first variable "time"
//and consider this a 1D histogram
// see https://root.cern.ch/root/htmldoc/guides/users-guide/Histograms.html
void prob15_1minuit() {
//Load the data and put it into a histogram
TH1D* hist = new TH1D("hist","",10,304.9,754.9);
ifstream infile;
infile.open("prob15_1.dat");
string line;
```

```
long int j=0;
if (infile.is_open()) {
while ( getline(infile, line)  ) {
istringstream ss(line);
ss » timetemp » xtemp;
x[j] = xtemp;
time[j]=timetemp;
//  z[j] =A*exp(-lam*time)+B;
j++;
hist->Fill(timetemp,xtemp);
//fills the histogram bins
}
std::cout « j « " entries loaded from file!" « std::endl;
}
infile.close();
//Declare the TMinuit object and set the function to minimize
TMinuit* myMinuit = new TMinuit(3); //3 parameters
myMinuit->SetFCN(FCN);
//Define the initial parameter guesses and step sizes
//  Double_t paramStart[2] = {1.8, 0.2};
Double_t paramStart[3] = {10000000.,0.004,85000.};
Double_t paramStep[3] = {100000.,.001,10000.};
//Declare the error flag and the parameters for Minuit to minimize over
//The 0.0, 0.0 parameters make it so the parameters have no limits
Int_t errFlag = 0;
myMinuit->mnparm(0,"A",paramStart[0],paramStep[0], 0.0,0.0, errFlag);
myMinuit->mnparm(1,"Lambda",paramStart[1],paramStep[1], 0.0,0.0,
errFlag);
myMinuit->mnparm(2,"B",paramStart[2],paramStep[2], 0.0,0.0, errFlag);
//Declare the minimization parameters
Double_t argList[3];
argList[0] = 500;      //Maximum number of iterations
argList[1] = 1.0;      //Functional tolerance
//The value of χ² − χ²min, which is used for the limit, can
// be reset by resetting ERR.
// The default value for one standard deviation is 1
//for χ² fitting and  should be 0.5 for maximum likelihood fitting
// because the log likelihood contains χ²/2.
// In general the parameter can be set to look at two or three standard
// deviation limits or 90% confidence levels and so on.)
//For χ², setting the parameter four gives two standard deviations,
//nine gives three standard deviations, i.e., it goes as the
//square of the number of standard deviations desired.
// Use statements below to SET ERR to .5 for log max.
//likelihood fits
```

```
//————————————————————
//  argList[0] = .5;  // error level for maximum likelihood minimization
//  myMinuit->mnexcm("SET ERR", argList, 1, errFlag);
// SET ERR=.5 only takes effect with  HESSE;
//the errors from MIGRAD are for 1.0 and are sigma are sqrt(2) larger
//  argList[0] = 500; // reset as argList used in MIGRAD and HESSE
//————————————————————
//Do the minimization with MIGRAD
myMinuit->mnexcm("MIGRAD", argList, 3, errFlag);
//Get the minimization results
double Avalue, AError, LambdaValue, LambdaError, BValue, BError;
myMinuit->GetParameter(0,Avalue, AError);
myMinuit->GetParameter(1,LambdaValue, LambdaError);
myMinuit->GetParameter(2,BValue,BError);
//  Double_t bValue, bError, mValue, mError;
//  myMinuit->GetParameter(1, mValue, mError);
Double_t funcMin, funcErrDist, errDef;
Int_t nAvPar, nExPar, iStat;
myMinuit->mnstat(funcMin, funcErrDist, errDef, nAvPar, nExPar, iStat);
myMinuit->mnexcm("HESSE", argList, 3, errFlag);
std::cout « "A" «Avalue « "+/-" « AError « std::endl;
std::cout « "Lambda" «LambdaValue « "+/-" « LambdaError «std::endl;
std::cout « "B" « BValue « "+/-" « BError „std::endl;
//These next lines are for making histograms
//To get the fitted curve we define the formula for the fit function
// and the range in x
TF1* fitFunc = new TF1("fitFunc", "[0]*exp(-[1]*x)+[2]",304.9,754.9);
fitFunc->SetParameters(Avalue, LambdaValue, BValue, 10.0*50.);
//Multiply by the bin width (50.)
hist->Draw();//This draws the histogram
fitFunc->Draw("L SAME");//This draws the fit to the histogram
}
```
(Fig. A.2)

MIGRAD is one of the Fit optimization methods. HESSE calculates the errors of the parameters. The output of this program (needed for Exercise 14.1)

10 entries loaded from file!

PARAMETER DEFINITIONS:

NO. NAME	VALUE	STEP SIZE	LIMITS
1 A	1.00000e+07	1.00000e+05	no limits
2 Lambda	4.00000e-03	1.00000e-03	no limits
3 B	8.50000e+04	1.00000e+04	no limits

Warning: Automatic variable ERRORDEF is allocated prob15_1minuit.c:77:
****

** 1 **MIGRAD   500   1 6.953e-310
****

FIRST CALL TO USER FUNCTION AT NEW START POINT, WITH IFLAG=4.
START MIGRAD MINIMIZATION. STRATEGY 1. CONVERGENCE WHEN
EDM .LT. 1.00e-03
FCN=2.47649e+07 FROM MIGRAD   STATUS=INITIATE   14 CALLS
15 TOTAL
EDM= unknown   STRATEGY= 1   NO ERROR MATRIX
EXT PARAMETER   CURRENT GUESS   STEP   FIRST
NO. NAME   VALUE   ERROR   SIZE   DERIVATIVE
 1 A   1.00000e+07 1.00000e+05 1.00000e+05 5.75265e+00
 2 Lambda   4.00000e-03 1.00000e-03 1.00000e-03 -2.77572e+10
 3 B   8.50000e+04 1.00000e+04 1.00000e+04 4.59228e+01
MIGRAD MINIMIZATION HAS CONVERGED.
MIGRAD WILL VERIFY CONVERGENCE AND ERROR MATRIX.
COVARIANCE MATRIX CALCULATED SUCCESSFULLY
FCN=69.7112 FROM MIGRAD   STATUS=CONVERGED   164 CALLS
165 TOTAL
EDM=0.000101955 STRATEGY= 1   ERROR MATRIX ACCURATE
EXT PARAMETER   STEP   FIRST
NO. NAME   VALUE   ERROR   SIZE   DERIVATIVE
 1 A   6.32421e+06 4.52258e+04 1.75050e+01 -1.44077e-06
 2 Lambda   7.02928e-03 2.21427e-05 1.01404e-08 2.61409e+03
 3 B   6.09705e+04 5.52756e+02 7.98635e-01 1.20033e-05
EXTERNAL ERROR MATRIX. NDIM= 25 NPAR= 3 ERR DEF=1
2.045e+09 9.842e-01 2.030e+07
9.842e-01 4.903e-10 1.082e-02
2.030e+07 1.082e-02 3.055e+05
PARAMETER CORRELATION COEFFICIENTS
NO. GLOBAL   1   2   3
1 0.99025 1.000 0.983 0.812
2 0.99376 0.983 1.000 0.884
3 0.93597 0.812 0.884 1.000
****
** 2 **HESSE   500   1 6.953e-310
****
COVARIANCE MATRIX CALCULATED SUCCESSFULLY
FCN=69.7112 FROM HESSE   STATUS=OK   16 CALLS   181
TOTAL
EDM=0.000101955 STRATEGY= 1   ERROR MATRIX ACCURATE
EXT PARAMETER `   INTERNAL   INTERNAL
NO. NAME   VALUE   ERROR   STEP SIZE   VALUE
 1 A   6.32421e+06 4.52353e+04 3.50099e+00 6.32421e+06
 2 Lambda   7.02928e-03 2.21474e-05 2.02808e-09 7.02928e-03
 3 B   6.09705e+04 5.52847e+02 1.59727e-01 6.09705e+04
EXTERNAL ERROR MATRIX. NDIM= 25 NPAR= 3 ERR DEF=1
2.046e+09 9.846e-01 2.031e+07

9.846e-01 4.905e-10 1.083e-02
. 2.031e+07 1.083e-02 3.056e+05
PARAMETER CORRELATION COEFFICIENTS
NO. GLOBAL   1   2   3
1 0.99025 1.000 0.983 0.812
2 0.99376 0.983 1.000 0.884
3 0.93599 0.812 0.884 1.000
A6.32421e+06+/-45225.8
Lambda0.00702928+/-2.21427e-05
B60970.5+/-

Fig. A.2 Figure from the MINUIT run. Here one compares the original histogram and the fitted curve

Appendix B
Obtaining Pseudorandom Numbers in C++

Use the following include files:

```
#include <iostream>
#include <random>
#include <fstream>
#include <math.h
#include <cmath> // allows access to functions like log and exp
#include <vector> // needed to use vectors
#include <random> // random number generation
#include <cstdlib> // needed for srand
std::random_device rd;
std::default_random_engine generator;
//If you wish to set an initial seed:
srand(1234)//or whatever unsigned number you want. The default is 1
std::uniform_real_distribution<double> distribution(0.0, 1.0);
//This generates a uniformly distributed set of random numbers from 0-1.
//If, instead you want a normally distributed set of random numbers
// with mean 0 and variance 1 then use:
std::normal_distribution<double> distribution(0.0,1.0);
// To use this:
x = distribution(rd); //gives x as an appropriate random number
//Use rd as the argument of distribution.
```

© Springer Nature Switzerland AG 2020
B. P. Roe, *Probability and Statistics in the Physical Sciences*,
Undergraduate Texts in Physics,
https://doi.org/10.1007/978-3-030-53694-7

Appendix C
Some Useful Spline Functions

An excellent set of programs for handling bspline problems is given
in the open source Gnu Scientific Library (GSL). A brief description
is:
int
gsl_bspline_workspace *bw;
// Allocate space for a bspline workspace. The size of the
//workspace is O(5k + nbreak)
//   bw=gsl_bspline_alloc (const size_t k, const size_t nbreak)
where inputs are: k   - spline order (cubic = 4)
nbreak - number of breakpoints, so:
bw=gsl_bspline_alloc (k, nbreak);
// The size of the workspace is O($k^2 + 5k$+nbreak)
//Return: pointer to workspace
*/

---

int
gsl_bspline_knots (const gsl_vector * breakpts, bw);
//gsl_bspline_knots()
// Compute the knots from the given breakpoints:
//   knots(1:k) = breakpts(1)
//   knots(k+1:k+l-1) = breakpts(i), i = 2 .. l
//   knots(n+1:n+k) = breakpts(l + 1)
//where l is the number of polynomial pieces (l = nbreak - 1) and
//   n = k + l - 1
//(using matlab syntax for the arrays)
//The repeated knots at the beginning and end of the interval
//correspond to the continuity condition there. See Chapter 18.
//of [1].
//Inputs: breakpts - breakpoints

© Springer Nature Switzerland AG 2020
B. P. Roe, *Probability and Statistics in the Physical Sciences*,
Undergraduate Texts in Physics,
https://doi.org/10.1007/978-3-030-53694-7

```
//      bw       - bspline workspace
//Return: success or error
*/
```

```
/*
int
gsl_bspline_knots_uniform (const double a,b,bw);
gsl_bspline_knots_uniform()
// Construct uniformly spaced knots on the interval [a,b] using
//the previously specified number of breakpoints. 'a' is the position
//of the first breakpoint and 'b' is the position of the last
//breakpoint.
//Inputs: a - left side of interval
//      b - right side of interval
//      bw - bspline workspace
//Return: success or error
//Notes: 1) w->knots is modified to contain the uniformly spaced
//      knots
//      2) The knots vector is set up as follows (using octave syntax):
//      knots(1:k) = a
//      knots(k+1:k+l-1) = a + i*delta, i = 1 .. l - 1
//      knots(n+1:n+k) = b
*/
```

```
int
gsl_bspline_eval (const double x, gsl_vector * B,bw);
/*
//gsl_bspline_eval()
// Evaluate the basis functions B_i(x) for all i. This is
//a wrapper function for gsl_bspline_eval_nonzero() which
//formats the output in a nice way.
//Inputs: x - point for evaluation
//      B - (output) where to store B_i(x) values
//          the length of this vector is
//          n = nbreak + k - 2 = l + k - 1 = w->n
//      bw - bspline workspace
//Return: success or error
//Notes: The w->knots vector must be initialized prior to calling
//      this function (see gsl_bspline_knots())
*/
```

```
int
gsl_bspline_eval_nonzero (const double x, gsl_vector * Bk, size_t * istart,
size_t * iend,bw);
/*
```

```
//gsl_bspline_eval_nonzero()
// Evaluate all non-zero B-spline functions at point x.
//These are the B_i(x) for i in [istart, iend].
//Always B_i(x) = 0 for i < istart and for i > iend.
//Inputs: x     - point at which to evaluate splines
//      Bk    - (output) where to store B-spline values (length k)
//      istart - (output) B-spline function index of
//             first non-zero basis for given x
//      iend   - (output) B-spline function index of
//             last non-zero basis for given x.
//             This is also the knot index corresponding to x.
//      bw    - bspline workspace
//Return: success or error
//Notes: 1) the w->knots vector must be initialized before calling
//      this function
//      2) On output, B contains
//        [B_istart,k, B_istart+1,k,
//        ..., B_iend-1,k, B_iend,k]
//      evaluated at the given x.
*/
```

```
int
gsl_bspline_deriv_eval (const double x, const size_t nderiv,
gsl_matrix * dB, bw);
/*
//gsl_bspline_deriv_eval()
// Evaluate dĵ/dxĵ B_i(x) for all i, 0 <= j <= nderiv.
//This is a wrapper function for gsl_bspline_deriv_eval_nonzero()
//which formats the output in a nice way.
//Inputs: x     - point for evaluation
//      nderiv - number of derivatives to compute, inclusive.
//      dB    - (output) where to store dĵ/dxĵ B_i(x)
//             values. the size of this matrix is
//             (n = nbreak + k - 2 = 1 + k - 1 = w->n)
//             by (nderiv + 1)
//      bw    - bspline derivative workspace
//Return: success or error
//Notes: 1) The w->knots vector must be initialized prior to calling
//      this function (see gsl_bspline_knots())
//      2) based on PPPACK's bsplvd
*/
```

```
int
gsl_bspline_deriv_eval_nonzero (const double x, const size_t nderiv,
gsl_matrix * dB, size_t * istart,
```

```
size_t * iend, bw);
//gsl_bspline_deriv_eval_nonzero()
```

// At point x evaluate all requested, non-zero B-spline function
//derivatives and store them in dB. These are the
//dĵ/dxĵ B_i(x) with i in [istart, iend] and j in [0, nderiv].
//Always dĵ/dxĵ B_i(x) = 0 for i < istart and for i > iend.
//Inputs: x   - point at which to evaluate splines
//   nderiv - number of derivatives to request, inclusive
//   dB   - (output) where to store dB-spline derivatives
//   (size k by nderiv + 1)
//   istart - (output) B-spline function index of
//   first non-zero basis for given x
//   iend   - (output) B-spline function index of
//   last non-zero basis for given x.
//   This is also the knot index corresponding to x.
//   bw   - bspline derivative workspace
//Return: success or error
//Notes: 1) the w->knots vector must be initialized before calling
//   this function
//   2) On output, dB contains
//   [[B_istart, k, ..., dn̂deriv/dxn̂deriv B_istart ,k],
//   [B_istart+1,k, ..., dn̂deriv/dxn̂deriv B_istart+1,k],
//   ...
//   [B_iend-1, k, ..., dn̂deriv/dxn̂deriv B_iend-1, k],
//   [B_iend,   k, ..., dn̂deriv/dxn̂deriv B_iend,   k]]
//   evaluated at x. B_istart, k is stored in dB(0,0).
//   Each additional column contains an additional derivative.
//   3) Note that the zero-th column of the result contains the
//   0th derivative, which is simply a function evaluation.
//   4) based on PPPACK's bsplvd
*/

// After downloading
// the program set, the invocations are (if GSL is in opt/local:
// cc -Wall -I/opt/local/include -c bsplineexample.c
// This produces a bsplineexample.o file and then follow this with;
// gcc -L/opt/local/lib bsplineexample.o -lgsl -lgslcblas -lm
// which produces the a.out file. A sample program is:
```c
#include <stdio.h>
#include <stdlib.h>
#include <math.h>
#include <gsl/gsl_bspline.h>
#include <gsl/gsl_multifit.h>
#include <gsl/gsl_rng.h>
#include <gsl/gsl_randist.h>
```

```c
#include <gsl/gsl_statistics.h>
/* number of data points to fit */
#define N      200;
// on some compilers the ";" may not be required for #define statements
/* number of fit coefficients */
#define NCOEFFS  12;
/* nbreak = ncoeffs + 2 - k = ncoeffs - 2 since k = 4 */
#define NBREAK   (NCOEFFS - 2);
int
main (void)
{
const size_t n = N;
const size_t ncoeffs = NCOEFFS;
const size_t nbreak = NBREAK;
size_t i, j;
gsl_bspline_workspace *bw;
gsl_vector *B;
double dy;
gsl_rng *r;
gsl_vector *c, *w;
gsl_vector *x, *y;
gsl_matrix *X, *cov;
gsl_multifit_linear_workspace *mw;
double chisq, Rsq, dof, tss;
gsl_rng_env_setup();
r = gsl_rng_alloc(gsl_rng_default);
/* allocate a cubic bspline workspace (k = 4) */
bw = gsl_bspline_alloc(4, nbreak);
B = gsl_vector_alloc(ncoeffs);
x = gsl_vector_alloc(n);
y = gsl_vector_alloc(n);
X = gsl_matrix_alloc(n, ncoeffs);
c = gsl_vector_alloc(ncoeffs);
w = gsl_vector_alloc(n);
cov = gsl_matrix_alloc(ncoeffs, ncoeffs);
mw = gsl_multifit_linear_alloc(n, ncoeffs);
/* this is the data to be fitted */
for (i = 0; i < n; ++i)
{
double sigma;
double xi = (15.0 / (N - 1)) * i;
double yi = cos(xi) * exp(-0.1 * xi);
sigma = 0.1 * yi;
dy = gsl_ran_gaussian(r, sigma);
yi += dy;
```

```
gsl_vector_set(x, i, xi);
gsl_vector_set(y, i, yi);
gsl_vector_set(w, i, 1.0 / (sigma * sigma));
printf("%f %f \n", xi, yi);
}
/* use uniform breakpoints on [0, 15] */
gsl_bspline_knots_uniform(0.0, 15.0, bw);
/* construct the fit matrix X */
for (i = 0; i < n; ++i)
{
double xi = gsl_vector_get(x, i);
/* compute B_j(xi) for all j */
gsl_bspline_eval(xi, B, bw);
/* fill in row i of X */
for (j = 0; j < ncoeffs; ++j)
{
double Bj = gsl_vector_get(B, j);
gsl_matrix_set(X, i, j, Bj);
}
}
/* do the fit  for bsplines to knots*/
gsl_multifit_wlinear(X, w, y, c, cov, &chisq, mw);
dof = n - ncoeffs;
tss = gsl_stats_wtss(w->data, 1, y->data, 1, y->size);
Rsq = 1.0 - chisq / tss;
fprintf(stderr, "chisq/dof = %e, Rsq = %f\n",
chisq / dof, Rsq);
printf(" n n");
/* output the smoothed curve */
{
double xi, yi, yerr;
for (xi = 0.0; xi < 15.0; xi += 0.1)
{
gsl_bspline_eval(xi, B, bw);
gsl_multifit_linear_est(B, c, cov, &yi, &yerr);
printf("%f %f\n", xi, yi);
}
}
gsl_rng_free(r);
gsl_bspline_free(bw);
gsl_vector_free(B);
gsl_vector_free(x);
gsl_vector_free(y);
gsl_matrix_free(X);
gsl_vector_free(c);
```

```
gsl_vector_free(w);
gsl_matrix_free(cov);
gsl_multifit_linear_free(mw);
return 0;
} /* main() */
```

Index

A

Acceptance rejection method, 74, 75, 82
A posteriori probability, 3
A priori probability, 1, 3
Arcsine distribution, 92
Arcsine law, 91
Arrangements, 28, 37
Asymptotically normal distribution, 123, 223
Asymptotic efficiency of an estimate, 122, 123

B

Background, 12, 26, 41, 80, 167, 181, 191, 192, 194, 198, 228, 234
Background subtraction, 26, 181, 197, 198, 228
Bartlett S function, 142, 215, 216, 223, 224
Bayesian approach, 171
Bayes theorem, 145–148, 152, 155, 158, 168
Bernoulli trials, 33, 40, 65, 66, 113, 114
Biased estimate, 141
Binomial coefficient, 28, 37
Binomial distribution, 33
Birth and death problem, 88
Blood test, 146
Boosted decision trees
  AdaBoost, 258, 259
  boosted decision trees, 256
  boosting, 258
  classifiers, 259
  decision trees, 257
  ϵBoost, 259
  Gini criterion, 257
  nodes, branches, and leaves, 257
  χ statistic, 259

Bridge hands of cards, 28, 30, 37
Brownian motion, 16, 17
Bubble chamber, 141
Buffon's Needle, 12

C

Central limit theorem, 43, 51, 66, 68, 107, 109, 111, 113–115
Central moments, 7, 12, 25
CERN, 80, 189
Change of variables, 18, 21, 206, 223
Characteristic functions, 61, 66–68, 100, 103, 104
Chauvenet's criterion, 92
Cofactor, 102, 196
Coin toss, 1–3, 33, 241
Combinatorials, 27, 29
Combining probability estimates, 210
Combining several measurements, 22, 23, 25, 140, 243
Composition, 61, 72
Compound probability, 61, 62, 68
Computer program, 2, 12, 80, 82
Computing, 12, 71, 75, 78–80, 108, 115, 172, 173, 198, 243
Conditional coverage, 160
Conditional density, 159
Conditional probability, 6, 36, 97, 145
Confidence belt, 150
Confidence interval, 145, 148–154
Confidence level, 149, 152, 153, 170–172, 180
Confidence limits, 145, 149, 151, 168, 172, 220
Confidence regions, 155, 191

© Springer Nature Switzerland AG 2020
B. P. Roe, *Probability and Statistics in the Physical Sciences*,
Undergraduate Texts in Physics,
https://doi.org/10.1007/978-3-030-53694-7